Central and Eastern European Development Studies

Springer
Berlin
Heidelberg
New York
Barcelona
Budapest
Hong Kong
London
Milan
Paris
Santa Clara
Singapore
Tokyo

Ulrich Graute (Ed.)
Institute of Ecological and Regional Development/
Institut für ökologische Raumentwicklung e.V.

Sustainable Development for Central and Eastern Europe

Spatial Development in the European Context

With 41 Figures and 28 Tables

Springer

Volume Editor

Ulrich Graute, M.A.
Institute of Ecological and Regional Development
Weberplatz 1
D-01217 Dresden
Germany

Cataloging-in-Publication Data applied for

Die Deutsche Bibliothek – CIP-Einheitsaufnahme

Sustainable development for Central and Eastern Europe: spatial development in the European context; with 28 tables / Ulrich Graute (ed.). – Berlin; Heidelberg; New York; Barcelona; Budapest; Hong Kong; London; Milan; Paris; Santa Clara; Singapore; Tokyo: Springer, 1998
(Central and Eastern European development studies)
ISBN 3-540-63814-8

ISBN 3-540-63814-8 Springer-Verlag Berlin Heidelberg New York

Typesetting: Camera ready copy by the editor
Cover design: Erich Kirchner, Heidelberg

SPIN 10569658 30/3136 – 5 4 3 2 1 0 – Printed on acid-free paper

Foreword

Central and Eastern European Countries are undergoing a difficult transition process with consequences for spatial development. Europe needs a more balanced, polycentric system of cities, a new urban-rural relationship, parity of access to infrastructure and knowledge as well as prudent management and development of the natural and cultural heritage.

Dealing with these complex challenges and the need for sustainable development politicians requires qualified advice from experts. To this end, the German Ministry for Regional Planning, Construction and Urban Development initiated the Network of Spatial Research Institutes in Central and Eastern Europe (CEE Net) in 1992. Exchange of knowledge and an open dialogue at all levels are preconditions for the cooperative development of visions and strategies serving spatial development in Europe. The annual conferences of the network have been a first steps towards closer cooperation. The high interest shown in these conferences by academics and government representatives and the fact that several research projects have already crystallised out of the debates they engendered indicate that the network is meeting a need for exchange and cooperation.

Decisions on spatial development do not stop at national borders and the discussion on spatial development in any European region has therefore to take the European context into consideration. A special advantage of this publication is that, as well as focusing on spatial development and research in CEE regions, it also discusses Central and Eastern Europe in a European context. The draft of the European Spatial Development Perspective (ESDP) presented in June 1997 to the Informal Council of EU Ministers responsible for spatial planning at Noordwijk, and which is discussed in this book, contains an exhortation to intensify cooperation with CEE partners. I hope this book will be a starting point for such intensified cooperation.

Finally, I would like to express my appreciation of the fact that two east German research institutes have taken a lead by commencing publication of the series 'Central and Eastern European Development Studies'. The series is getting underway at about the same time as the Community Initiative INTERREG II C on transnational cooperation in spatial development, and the two could well prove to be complementary. Projects from the Baltic Sea region and the Central, Adriatic, Danubian and South-East European Space of INTERREG II C are likely to be discussed in the books of the series and this discussion could in turn be important for the further evolution of INTERREG II C and of cooperation on spatial development in Europe.

Bonn, August 1997

Professor Dr. Klaus Töpfer
Federal Minister for Regional Planning,
Construction and Urban Development

In November 1997 Klaus Töpfer was nominated by the United Nations Secretary-General, Kofi Annan, as Executive Director of the United Nations Environment Programme (UNEP).

Foreword

Preface to the Series

With the presentation of the first official draft for a European Spatial Development Perspective, the contours of a future European spatial development policy have become clearer. Its fleshing-out and implementation will be one of the most pressing items of trans-European debate over the years to come. Parties to this debate will be not only the member states of the European Union but also and in particular the bordering states in central and eastern Europe, who even now are having a hand, within the framework of transnational schemes, in elaborating a European agenda and putting it to effect.

One of the most momentous of the schemes is the INTERREG II C Community Initiative, which is explicitly geared towards supporting transnational operational programmes in the sphere of spatial planning in specified cooperation zones such as, for instance, the Baltic Sea Region and the Central, Adriatic, Danubian and South-Eastern European Space. There are five clusters of measures based on observations and recommendations contained in the European Spatial Development Perspective that will acquire special significance here. They are:

- Initiatives for the launching of a cooperative spatial planning process;
- Promoting cooperation for the establishment of more balanced and polycentric urban systems and settlement patterns;
- Development of multimodality in transport systems and parity of access to infrastructures;
- Cooperation for the improving of access to knowledge and information;
- Prudent management and development of natural and cultural heritages.

It is thus a very opportune moment to be introducing a series that provides a forum for transnational exchanges of views and discussion of the latest scientific findings on spatial development for a segment of Europe that is currently undergoing a turbulent process of transition. Entitled "Central and Eastern European Development Studies", the series will be intent on fulfilling this brief. There will be three main areas of focus:
Firstly, spatial development in central and eastern European regions. Included here will be regional studies, studies on regionalisation processes in central and eastern Europe, and comparative investigations into individual aspects of spatial development in central and eastern Europe.

Secondly, the evolution of central and eastern Europe in the context of Europe as a whole. This area of focus will concern itself with European integration and spatial development in central and eastern Europe, the influence of EU policy and in particular of the EU's spatial development policy on central and eastern Europe, and development ties between central and eastern Europe and bordering regions.

Thirdly, spatial science in central and eastern Europe. This will primarily involve contributions from central and eastern Europe on the development of spatial science research (theory formation, methods, tools etc.) as well as studies that further dialogue and cooperation in spatial science.

The current first volume of the new series grapples with one of the most crucial issues of the present time, namely the options available for shaping sustainable development. It is the upshot of a scientific exchange conducted within the framework of a research network for the central and eastern European states whose secretariat is accommodated in the Institute of Ecological and Regional Development (IÖR) in Dresden and has been headed since its inception by the editor of this book, Mr Ulrich Graute. The contributions contained within the current volume are the result of several years of research whilst at the same time embracing up-to-the-minute developments in spatial planning as well as topical issues of spatial development policy. The series will in future serve as a forum for all those actively taking part in the debate on changes in the central and eastern European space and helping fashion them.

Dresden and Rostock, Autumn 1997

Professor Dr. Wilfried Erbguth
Rostock University
Baltic Sea Institute for Marine
and Environmental Law

Professor Dr. Bernhard Müller
Institute of Ecological
and Regional Development,
Dresden

Preface

This book, the first in a series, provides insights into scientific studies and political discussion on spatial development in central and eastern Europe. 'Sustainable Development for Central and Eastern Europe' with its subtitle 'Spatial Development in the European Context' is a fairly well-defined subject yet one that is almost too complex to be covered in a single volume.

The subject is well-defined in so far as the focus is on *Central and Eastern Europe (CEE)*, but what does this specifically mean? What actually constitutes CEE can be interpreted in various ways, and in some respects it is even practical to keep the definition open. Of course, when officials from the European Union, and indeed many others, speak about CEE countries they are usually referring to the countries of the former Council of Mutual Economic Aid (CMEA), but these countries no longer form a 'bloc' and there are considerable differences in the way they are developing politically and economically. From a geographical point of view, it is wrong to speak of central Europe without including countries such as Austria and Germany. In the context of cross-border or transnational cooperation in spatial development, it has become established practice to consider countries further to the north (Scandinavia) and to the south (Italy, Greece, and Turkey) together with CEE countries. For a multidisciplinary subject such as spatial development and especially when discussion of it is specifically focused on cross-border or transnational issues, a narrow definition of CEE can be dispensed with. Instead, when it comes to outlining development perspectives for the central and eastern European 'space', this needs to be done with an eye to the wider European context and processes of pan-European integration.

The number of issues relevant to sustainable development is practically infinite. This book is accordingly confining itself to the most salient aspects of European spatial development. *Spatial development* is understood as the process that gives geographical expression to the economic, social, cultural and ecological policies of society. It is at one and the same time a scientific discipline, an administrative technique, and a policy devised as an interdisciplinary and comprehensive means of achieving balanced development and the physical organisation of space within the framework of an overall strategy. *European spatial development* is understood to be that aspect of spatial development that transcends national considerations. It necessarily embraces visions, guiding principles (*Leitbilder*), and development schemes for the European space as well as the instruments for their implementation. The specific purpose of spatial development schemes is to generate an outline framework for urban and regional planning as a means of preventing uncoordinated and short-sighted development activity, and it is precisely the narrowness of its remit that makes spatial development such a complex field of research and development. It is a field in which everything treated discursively – with regard to, for example, sustainable development – also needs to be rendered physically visible. There are practically as many issues with a bearing on spatial development as there are with a bearing on sustainable development. Inevitably, a scientific book in which two such complex subjects are discussed together, and with reference to an area covering about half of geographical Europe, will be open to the charge of having omitted this or that important aspect. That the venture has nonetheless been proceeded with has to do with the need to learn how to cope with these complex challenges. As well as encapsulating the present state of thinking within the Network of Spatial Research Institutes in Central and Eastern Europe (CEE Net), it is to be hoped this book will stimulate further discussion and, with it, research projects.

Contributions are divided into two main sections, with the issues dealt with in the main body of the book already being given a substantial airing in the editor's introduction, 'The European Debate on Sustainable Spatial Development', which, as well as constituting a general introduction to the subject-matter of the book, also addresses itself to broader discourses on sustainability and European integration.

Given the significance to the development process in central and eastern Europe of the broader European context, this strand of discussion is documented in the first principal section of the book (*PART I*). The Council of Europe (see Tarcisio Bassi) instigated a debate on transnational development of a sustainable nature at a very early stage. Among the main players currently involved in elaborating a European Spatial Development Policy within the EU are the national governments of Germany (see Welf Selke) and of the Netherlands (see Karolijn van Ginneken).

The second principal section (*Part II*), which forms the bulk of the book, focuses on efforts towards sustainable development in selected regions of central and eastern Europe. The regions selected represent major challenges for the CEE countries, and the authors of the contributions describe the respective regional efforts of researchers and planners to cope with these challenges.

In the first group of articles cross-border cooperation and development at the external borders of the European Union are discussed. Viktor Freiherr von Malchus and Isolde Roch provide an overview of cross-border cooperation at the external borders of the EU that takes in both its history and its present state of development, while Wilfried Altzinger, Gunther Maier and Jarko Fidrmuc review predominantly economic aspects of cross-border development as evidenced in the emerging urban agglomeration of Vienna/Bratislava. Perttu Vartiainen focuses on cross-border urban networking as a learning process and in the process offers a number of insights into the problems of cross-border cooperation in practice.

The Danube is a river having to square its identities as a waterway and as a candidate for water conservation, and this is reflected in contributions by Peter Schneidewind, who discusses the river as a trans-European transport resource, and by Simona Pascariu, who focuses on water preservation strategies and the regional plan for the Danube Delta Biosphere Reserve.

A third group of articles reflects on regional European cooperation in the Baltic Sea Region. Jacek Zaucha sets forth the history and development of the cooperation programme 'Vision and Strategies around the Baltic Sea 2010' (VASAB 2010), while Helle Fischer deals with coastal zone management, one of the main planks of sustainable spatial development in the Baltic Sea region.

Levels of cooperation in the Black Sea region are less evolved than in the case of the Baltic Sea. Accordingly, it is only possible to record points of departure and initial moves towards cooperation here. Alexander Shilov studies the ecological prospects of the Black Sea, while Vladimir Anderson expounds on his work in Odessa on devising a regional GIS for development planning along the Ukrainian coast. Among the few examples of cooperation in the Black Sea region featuring a spatial development dimension is the integrated coastal zone management scheme outlined by Igor Kharitonov.

A last group of articles seeks to analyse the impact on spatial development in Belarus and Ukraine of the catastrophe in the Chernobyl nuclear power plant. Irina Gukalova has written a piece on the socio-economic consequences of the Chernobyl catastrophe in Ukraine, while Valentin Yatsukhno and Ludmila Kozlovskaya have studied the disastrous ecological impact of the Chernobyl catastrophe on development in Belarus. Leonid Rudenko and Inga Horlenko provide additional information concerning the outlook for sustainable development in Ukraine. The last contribution in this group is by Victor Ivlichev and Isaac Khazdan, who demonstrate with their regional development concept for the Brest–Minsk–Orsha section of the trans-European transport corridor from Berlin to Moscow that, Chernobyl notwithstanding, there are other major challenges facing spatial development in the region.

The concluding article of the book entitled 'Towards a Framework for Sustainable Spatial Development in CEE' summarises perspectives for sustainable spatial development in central and eastern Europe.

Had it not been for the debate initiated within the Network of Spatial Research Institutes in Central and Eastern Europe (CEE Net) and, most notably, the conferences of the Network in Vilnius 1995 and Odessa 1996, this book could not have been written. As the editor of this book, I would therefore like to express my gratitude to the German Ministry for Regional Planning, Construction and Urban Development and to others who have actively supported the CEE Net. This book has also been made possible through the support of the *Institut für ökologische Raumentwicklung* (Institute of Ecological and Regional Development, IÖR) and its Director Professor Bernhard Müller. In addition, I wish to thank Gisela Richter of the IÖR for her desktop publishing exploits and Timothy J. S. Driver for English language editing and translation work.

Dresden, Autumn 1997 Ulrich Graute

Contents

The European Debate on Sustainable Spatial Development

Ulrich Graute
Institut für ökologische Raumentwicklung (IÖR)
Institute of Ecological and Regional Development
Dresden, Germany

1 Introduction

While sustainability and sustainable development are major subjects of discussion not only among experts but also among the general public, spatial development by contrast is mostly a matter of concern for urban and regional planning experts, some scientists, and a minority lay audience. This is surprising in so far as the discussion on sustainability focuses on many 'new' aspects which are not that new in the context of spatial planning. For example, the report of the Brundtland Commission and the UN Conference in Rio de Janeiro stated that future development must meet the needs of the present without undermining the ability of future generations to satisfy their own needs. This need for a longer-term conception of development planning was central to the emergence of disciplines such as urban and regional planning early this century. Neither is the integrated, comprehensive approach espoused in the current debate on sustainability and calling for ecological, cost-effective, and socially balanced development actually very new to spatial development. The spatial planning (*Raumordnung*) policies forged in the industrial and metropolitan regions of Germany from the 1920s, for instance, were an answer to growing environmental and supply-related problems resulting from the uncoordinated development of industrial sites and settlements (Richter 1994). Unfortunately, spatial development at that time tended to focus on economic and social development and in many respects disregarded environmental issues. As environmental protection developed to become an important political and economic factor, however, spatial planners became more concerned about a truly integrated and comprehensive approach to development. Today, it is a matter of course that spatial development perspectives envisage development that is sustainable and that planners endeavour to integrate the principles of sustainable development into their planning schemes. It is one of the fascinating aspects of spatial development that it is in precisely this field that in the long run the vision of a sustainable form of development needs to assume physically discernible contours.

Five years after the Rio conference it has to be admitted that in spite of many advances sustainability and sustainable spatial development are still more a general model than a well-defined system merely needing to be put to effect. This state of affairs has particular implications for a region such as central and eastern Europe that is gradually integrating with other parts of the continent and is at the same time undergoing a difficult transition process.

Given the virtually infinite number of issues having a bearing on spatial development it is necessary for a book like this to be selective. In this article, for instance, the focus is on European spatial development cooperation. It is intended as an introduction to the subject of European spatial development but at the same time it points up a key aspect of spatial development, to wit the debate surrounding the issue. European spatial development is going to be an *ongoing process* (see van Ginneken) in which planning and implementation are interrelated. Planning and implementation will themselves *evolve* and this evolution will constitute a *learning process* for the various players involved (see Vartiainen). Hence, discussion, cooperation, and other forms of communication relating to sustainable spatial development are not just a side-issue but are rather part of the development process itself.

Discussions on European cooperation in spatial development mostly relate to the EU spatial development policy now gradually taking shape, even in many central and eastern European countries. There are two reasons for this. Firstly, within the Union and its member states the debate has a longer tradition than in central and eastern Europe (CEE) and basic models have already been constructed (e.g. ESDP). Secondly, the enlargement of the EU and the extension of the applicability of its policies to the East is considered inevitable by many experts in the East and has therefore to be addressed when considering development in CEE (cf. Savy et al 1996, p 43). Obviously, it is not possible at this juncture to state whether or when given CEE countries will be admitted to the European Union, but it should be noted that in the Baltic Sea region even countries very far to the East such as Russia and Belarus are participating in European spatial development activities organised in close cooperation with, and with the support of, the EU (see Zaucha, Ivlichev and Khazdan; also VASAB 1996, Graute 1996c).

2 European Spatial Development

2.1 Language and Terminology

For the purposes of this article, *spatial development* is understood as the process which gives geographical expression to the economic, social, cultural and ecological policies of society. It is at one and the same time a scientific discipline, an administrative technique, and a policy conceived as an interdisciplinary, all-embracing approach towards ensuring balanced development and the physical organisation of space in accordance with an overarching strategy. *European spatial development* by turn is understood as that aspect of spatial development which transcends the national dimension. It embraces visions, models, and development schemes for the European space as well as the instruments for their implementation. These definitions draw on similar explications propounded by the Council of Europe (1983) and by von Malchus (1995, pp 261–268).

Hitherto, there has existed no common language and no standardised terminology for the European debate on spatial development. Once the borders of the former eastern bloc countries had been opened up to the West, Russian lost its role as the *lingua franca* for international communication in the area. Of course, Russian is still widely understood, but even in the Baltic states and Ukraine it is greeted with much resentment, for the time being at least, because of its associations with the old oppressive empire. In most countries of the former Council of Mutual Economic Aid (CMEA) part of the process of wrenching oneself away from Moscow's clutches involves promoting the national language in preference to Russian. Besides, western languages such as English, German or French are of growing importance for international links with counterparts in Europe and the world. In the field of cooperation on spatial development there is a trend towards English as the principal language for communication among experts. Multilingual conferences in central and eastern Europe do not always offer German or Russian translations while English is always catered for. Nonetheless, English is not always needed as mediator. For example, at the external border of the EU in central Europe German is the main language for communication, and in spite of current reservations about Russian it is not unlikely that in the long run this language will again become the main vehicle for dialogue among experts in eastern Europe. The fact that there is no common language in central and eastern Europe is of importance for transnational cooperation in spatial planning in so far as, owing to the absence of a common language, there is no common terminology. With respect to the wider European context it may be added that the situation in western Europe is not that different, since France, Germany and other countries with a strong tradition in spatial planning have generated terminologies of their own in their own national languages and in some cases the terms are difficult to translate through direct renderings.

The English term ‘regional planning’, for example, can be translated into German as either ‘Regionalplanung’ or ‘Raumordnung’ (depending on its specific meaning), whilst ‘Raumordnung’ by turn is often translated into English as ‘spatial planning’ (cf. English version of Fürst et al 1994). The French ‘aménagement du territoire’ and the German ‘Raumordnung’ are not exact semantic equivalents but for both the English terms ‘spatial planning’ and ‘regional planning’ are current (ARL 1996, pp 3–4). The Council of Europe, probably with such terminological problems in mind, does not use the term spatial planning alone, but instead speaks of ‘regional/spatial planning’ (see Bassi). The definition of ‘European spatial planning’ (Europäische Raumordnung) by von Malchus in the German-language *ARL-Handbuch für Raumordnung* does not cover European regional planning. In fact, this term is not mentioned at all in the book. Instead, there is a separate article on European regional policy written by Treuner (1995, pp 268–271). According to Treuner, EU regional policy lacks the dimension of spatial planning. This distinction between regional policy and spatial planning in the European context is desirable since regional policy as hitherto practised by the European Union does not incorporate spatial planning. The fact that Treuner uses the term *European regional policy* and not European regional planning is another departure from the Council of Europe's terminology but the definition in the ARL-Handbook is again worth opting for given that practical implementation of European regional policy merely takes the form of action supporting regional planning by national or regional authorities. It would be wrong to claim that the EU or any other European institution is planning the development of regions, and without a European player taking responsibility for regional planning it is somewhat problematic to speak of 'European regional planning'.

A further inconsistency in the use of terms concerns the notions of *planning* and *development*. Planning has traditionally been understood to mean the process whereby development schemes are prepared and regulated prior to implementation, though this semantic division is gradually disappearing and it is now often the case that only 'development' is used. Not only does 'planning' have an often negative connotation as far as the general public is concerned, in practice it is becoming more and more difficult to differentiate between planning and development processes. Stating clearly where planning ends and development begins is particularly difficult at the transnational level. Neither for the European level, the Baltic Sea, nor any other European region does there exist an authoritative fully-fledged plan for spatial development that is merely awaiting implementation. What does exist is a variety of development schemes (see contributions by Zaucha and van Ginneken), but these are not and are not intended to be comprehensive mega-plans or end-products. Instead, their aim is to point the way for a coordinated mode of development and they will require updating and further refinement in line with changing conditions. Planning and development are increasingly interrelated processes with a long-term orientation and it is reasonable to consider planning more as part of the development process than as something separate from it.

At this point it might seem advisable to stick to the term 'European spatial development' and to steer clear of 'European regional planning', 'regional development', or 'spatial planning'. In the context of this article this is indeed what will be done, but it must be recognised that the mere act of eschewing the other formulations is not in itself sufficient to put an end to all the terminological contradictions that exist. The term 'region', for example, is not only used for the subnational level. Major European spaces such as the Black and the Baltic Sea areas are likewise called regions. This terminological doubling-up in English language causes problems in the arena of academic discussion. On the one hand it is very unlikely that a regional development scheme for these regions can be anything but a spatial development scheme and on the other hand it would be linguistically logical to call a spatial development scheme for a region a 'regional development scheme'. In conclusion it must be stated that *the practical use of terms like 'region', 'space', 'planning' and 'development' in public and scientific discussion is subject to inconsistencies, giving rise to misunderstandings that can actually impede transnational cooperation.*

Fortunately, an improvement is in sight. A more consistent terminology may well come into being as a side-product of the European Spatial Development Perspective (ESDP). To outline and discuss ESDP it is necessary to develop a common understanding of terms. The process of drafting and subsequently discussing ESDP thus represents a good opportunity to improve and standardise terminology in the field.

2.2 Necessity of Cooperation in European Spatial Development

2.2.1 Global and European Challenges

The number of issues having a bearing on spatial development is virtually endless and as a result there are many challenges to be met along the road to cooperation. A great number of European territorial issues are discussed among researchers, planners and politicians that necessitate transnational cooperation on spatial development. According to ESDP, territorial issues are above all, but not solely, those that occur at the continental and transnational level. They also include issues manifesting themselves at regional or local level as well as at many different locations in the Union. It is the European dimension that instils the element of common interest. (ESDP 1997, II.A; also van Ginneken). Most of the contributors to this book make reference in one way or another to these challenges. Set out below are the headings into which perceived tasks in hand have been categorised (see also EC 1994, pp 140–141).

The two concepts under which all other factors are frequently subsumed are European integration and economic globalisation.

European Integration

Furthering the process of European integration, which has received a major boost from the transformation now underway in the former communist countries of central and eastern Europe, is a prime objective of European cooperation in spatial development (cf. van Ginneken, Selke and Zaucha). Spatially significant decisions do not cease to be effective at national boundaries. Knowledge of what is going on in other, notably in neighbouring, countries and of imminent development schemes that might have a crucial bearing on a country's own development is indispensable (cf. Ivlichev and Khazdan on the TEN and developments in Belarus). The growing interdependence between the different parts of the European territory takes three forms: interdependence between territorial entities, that between the various sectoral policies of relevance to the territory, and that between the various levels of government sharing responsibility for a particular territorial unit (ESDP 1997).

Economic Globalisation
All planning systems and levels are being impacted upon by the Single European Market and the globalisation of economic activity and investment. Competition between states to attract inward investment is increasing, companies have become vastly more mobile, and states are having to cope with the incessant pressure associated with the demand for employment opportunities. This particularly affects cross-border development at the external borders of the EU (cf. Altzinger, Maier, and Fidrmuc).

While the two above-mentioned categories are still rather general in the way they impinge on spatial development, there is also the option of dividing the challenges involved into the following six categories.

Spatial Cohesion
Pursuing the ideal of cohesion involves efforts to tackle inter-regional disparities, problems and social conflicts relating to inter-regional economic development, and other elements of regional conflict whilst wherever possible preserving diversity. What is so far lacking is, in the words of the ESDP, a 'balanced European territory' (cf. van Ginneken).

Environmental Issues
Both residential and industrial development have always been conditioned by the availability and quality of environmental assets. With awareness of the finiteness of natural resources growing, environmental degradation on the increase, and at the same time demand for natural assets also expanding, environmental issues are becoming an ever more important factor for spatial development (cf. Fischer, Shilov, and Kharitonov). The six major environmental issues are waste reduction and management, climate change, loss of biodiversity, acidification and pesticides, forest degradation and freshwater management (ef. Roch and Pascariu).

Trans-European Networks
Energy supply, telecommunications, and transport are all subject to national planning, this despite the fact that the major networks have long since acquired a European dimension. In response to this, the EU is in the process of mapping out a series of Trans-European Networks (TEN). Degree of access to these networks will play a major part in defining a given region's future evolution (cf. Ivlichev, Khazdan, and Schneidewind).

Public Sector Financing
Public sector financing is increasingly becoming a key issue. In some countries this financing is provided primarily through local government and in others primarily through nationally-controlled public sector development agencies. With the EU Structural Funds scheme making the Union a major financier of regional and local development projects, the Union is acquiring an increasing say over

urban and regional development in those regions eligible under the scheme. How these funds are allocated is particularly important where the above-mentioned competition between states to attract inward investment is concerned. Through EU Community Initiatives such as PHARE, TACIS, and INTERREG and further funding schemes organised by the World Bank and the United Nations, international institutions are likewise gaining an increasing influence over spatial development in central and eastern Europe. The overall level of funding flowing into CEE countries may be low compared to that available for projects in western countries, but given the financial situation of many CEE countries these limited resources still have a comparatively significant impact where active development is feasible at all (cf. Kharitonov and Gukalova).

Efficiency of the Planning System
The ease with which planning permission is obtained may give a region or state a relative edge. Given a situation, for example, where there is ease of cross-border movements of goods, people, and money between two regions located at a common border yet they have different planning systems, that region with the more efficient planning system will be at an advantage (regarding obstacles to cross-border cooperation see Altzinger, Maier and Fidrmuc, and Vartiainen).

Population Developments
Major demographic and urbanisation trends are among the main forces necessitating European cooperation in spatial development. Patterns of migration have increased the severity and urgency of problems facing planners in large cities. Alongside migration within a country, immigration from beyond national borders also needs to be borne in mind. This represents a particular challenge for border regions, as is discussed in the contributions on cross-border development (see von Malchus, Vartiainen, Zaucha, Altzinger, Maier and Fidrmuc).

2.2.2 The Political Debate on a European Spatial Development Policy in Response to the Challenges Faced

The various challenges faced have given rise to a debate throughout the newly widened political European entity, but especially in western European countries such as Germany, on the necessity of European cooperation in spatial development and on how this cooperation is to be put in place and institutionalised. In keeping with the *principle of subsidiarity*, solutions to problems are to be sought at the lowest level possible. Only if a problem cannot be solved at local, regional, or national level is cooperation at international or transnational level foreseen. Spatial development falls under the principle of subsidiarity in the European Union (Article 3b EC Treaty, see also ARL 1996, p 48). There is no overall EU responsibility for spatial development planning. Urban and regional development are within the brief of local and regional administration (with the exception of centralised states where responsibility for regional development is vested in national government). Countries such as Austria, Germany, the Netherlands and France have strong traditions of spatial development and well-established administrative structures to that end. In most countries in eastern and central Europe development policies and administrative structures of this nature are either lacking or inchoate. So why and when would it be appropriate to contemplate cooperation in spataial development at at European level in such circumstances?

The fact that a spatial issue is capable of being solved through transnational cooperation is not reason enough on its own to adopt such a course of action. A necessary precondition is that proof is forthcoming that a solution could not be found at national or indeed lower administrative level. Only once this has been established can there be said to be a necessity for cross-border, transnational, or, where this is applicable, pan-European cooperation.

The *history* of the debate on European spatial development policy and its necessity stretches back almost fifty years. It began in the Council of Europe in 1949 and confined itself at that time to cross-border cooperation. The Council of Europe intensified its activities when the ministers responsible for spatial development first met in 1970, and in 1989 the equivalent ministers within what was now the EC started to meet on a regular (but informal) basis. INTERREG, the Community Initiative for cross-border cooperation, was launched in 1990 and in 1991 the EC Committee for Spatial Development (CSD) was set up. Since the conference at Rio de Janeiro in 1992 and as part of the intensifying process of European integration, there have been in-depth discussions on the need to strengthen cooperation on the issue of sustainable spatial development in Europe (see von Malchus, Bassi, Selke and van Ginneken).

In 1994, the *EU Commission* published its document 'Europe 2000+: Cooperation on European Territorial Development', which was the principal EU document relating to European spatial development until the draft of ESDP was published in June 1997. In the introduction to the *Europe 2000+* document, the nature of the need for European cooperation in spatial planning is expressed as follows: 'The basic theme which runs through *Europe 2000+* is that, as the European Union prepares to enter the 21st century, growing interdependence between the regions which make up the Union and the increasing relations with other parts of Europe call for closer and more systematic cooperation over territorial planning across Europe as whole. (...) It is the Commission's firm opinion that the Union's ultimate objective of sustained and balanced economic development can be achieved only through cooperation and joint action between all interested parties at all levels, supported by a common framework for territorial planning' (EC 1994, p 3). The fact that a supranational governmental institution in Europe publishes a document advocating greater cooperation at a European level may not be surprising, and it is questionable whether this document alone would suffice to lend concrete expression to the need for European cooperation in spatial development. It was crucial, therefore, that *Europe 2000+* was assented to by *the Informal Council of Ministers responsible for Spatial Planning* held in *Leipzig* on 21–22 September 1994. The first paragraph of the conclusions issued by the German Presidency states: 'All delegations hold the view that what is needed is a European spatial development perspective drawn up among the Member States themselves and with the participation of the Commission based on the strategy of sustainable development' (BMBau 1994, pp 17–18). The principle of subsidiarity appears in this phrase as a circumlocution: 'among the Member States themselves'. A more explicit reference to the principle was not necessary because the Commission had underlined its importance early on in the *Europe 2000+* document (EC 1994, p 3).

The *Treaty of Maastricht* does not include a section on spatial development but nevertheless articulates several objectives with a bearing on spatial development. The establishment of social and economic cohesion is one of the main pillars of the Union and its community policies (Title XIV, Articles 130a to 130e). A second goal of relevance to spatial development is the establishment of Trans-European Networks (TENs) in transport, telecommunications, and energy (Title XII, Articles 129b to 129d). The third related goal involves protection of the environment, including measures 'concerning town and country planning' and 'land use' (Title XVI, Articles 130r to 130t).

One of the reasons why a European spatial development perspective is needed ensues from the fact that the Treaty of Maastricht contains many spatially relevant sectoral policies but does not assign powers of authority in respect of spatial development policy. In pursuit of the goal of cohesion, aspects such as the creation of Trans-European Networks (TENs), cross-border cooperation, and the designation of 'community assets' warranting protection (e.g. a Europe-wide system of biotopes or the use of cross-border water systems) are particularly crucial to European spatial development and thus make it imperative that a suitable European policy be elaborated.

In September 1994 at Leipzig, the *Informal Council of Ministers responsible for Spatial Planning* in the member states of the EU adopted the principles for a European Spatial Development Policy. In this document the governments declared that future ESDP will adhere to the following principal tenets (BMBau 1995, p 11)

- spatial planning can contribute in a decisive way to the achievement of the goal of economic and social cohesion,
- the existing powers of the institutions responsible for Community policies remain unchanged; ESDP may contribute to the implementation of Community policies which have a territorial impact, but without constraining the responsible institutions in exercising their responsibilities,
- the central aim will be to achieve sustainable and balanced development,
- it will be prepared respecting existing institutions and will be non-binding on member states,
- it will respect the principle of subsidiarity,
- each country will take it forward according to the extent to which it wishes to take account of European spatial aspects in its national policies.

In November 1994, two months after these principles had been adopted, Bruce Millan, the EU Commissioner for Regional Policy, proposed that spatial development should be included in the revised Treaty of the Union.

Among the other EU bodies discussing European spatial development are the European Parliament and the Economic and Social Committee of the Union. For a number of years the *European Parliament* had a 'Commission for Regional Policy and the Development of the Territory', whereas in 1988 the European Parliament pointed out 'that Community regional policy came into being very late'. In the same text European spatial development makes a discreet appearance in a proposal suggesting that interventions made 'in the context of supraregional programmes are admissible if they are justified by regional planning policy' (quoted from Savy et al 1995, p 9).

In 1995 the *Economic and Social Committee of the European Union* approved the *Europe 2000+* document. It identifies 'a growing public awareness of the need for a spatial planning policy drawn up in consultation with all parties concerned - and, in particular, local and regional authorities and the socio-economic partners'. The opinion states: 'Further justification for the establishment of a European spatial planning policy is provided above all by the fact that spatial planning decisions are inextricably bound up with the implementation of other Community policies, because of the intermeshing and impact on spatial planning of the EU's mandatory policies' (quoted from Savy et al 1995, p 9).

In Germany *the Conference of Ministers responsible for spatial development (MKRO)* has proposed that the following paragraph should be added as a new paragraph (2) to Article 3 of the EC Treaty: 'The Community shall coordinate the policies of those departments whose activities have an impact on the development of the cities and regions of the Member States on the basis of the objectives for spatial development formulated by Member States for spatial development and a European Spatial Development Perspective (ESDP) to be drawn up in the process of cooperation among Member States'. A working group of the *German Academy of Spatial Research and Regional Planning (ARL)* on codification of European spatial development policy in the Treaty on the European Union supports the proposal of the MKRO and lists other possibilities for regulating the enactment of European spatial development policy: 'A case in point would be to amend Article 130a (1) or 130b EC Treaty, for example by adding the aspect of 'spatial cohesion', or by providing the instruments and procedures required by an ESDP and by instituting a duty to make periodic reports. A further recommendation would be the inclusion of a provision to supplement, for example, Art. 130a EC Treaty, under which the Community would be required to support institutional cooperation between local and regional territorial entities in connection with cross-border cooperation in areas along EU internal frontiers and in the proximity of the EU's external borders' (ARL 1996, p 49).

The discussion on European spatial development has been followed closely at regional level and in 1994 the *Assembly of European Regions* sent a questionnaire to its members to find out the regions' view of the territorial effects of European policies. 135 regions in 21 countries sent their replies to Strasbourg. The views of the regions which are summarised in a report presented by Robert Savy include some positive remarks on European spatial development cooperation but in their general approach, the regions are predominantly concerned with the diversity of Europe, which is to be preserved by conducting *ex ante* studies on the regional impacts of all community policies and by applying the principle of subsidiarity. This principle is emphasised in exploring means of achieving territorial cohesion, but the discussion of the principle ends with the conclusion 'that the principle of subsidiarity is not in conflict with the strengthening of territorial cohesion by Community policies' (Savy et al 1995, p 16 and p 140).

2.3 Sustainability and European Spatial Development

Sustainable development and the principle of subsidiarity are often discussed hand in hand. While sustainable development presupposes a comprehensive and integrated approach, the principle of subsidiarity holds that the decisions entailed by this approach are not to be made at the highest political and administrative level but to be decentralised wherever possible. The bottom-up approach of the subsidiarity principle effectively represents the exact opposite of what once obtained in the centrally planned systems of the East (top-down approach). Clearly, this fact will make it even more difficult for central and eastern Europe to achieve sustainable spatial development than the western countries.

When spatial development experts discuss sustainable spatial development, they occasionally begin by citing European documents on spatial development (e.g. *Europe 2000+*) but more often they allude to final documents such as the Declaration of the UN Conference at Rio in 1992. The conclusions and main documents of this conference not only gave a major impetus to the discussion on sustainability in general, they also contained avenues of thought of immense relevance to the discussion on transnational cooperation in spatial development. The Rio Declaration on Environment and Development (United Nations 1993) is a response to the challenges of globalisation and environmental pollution.

Nonetheless, the Rio documents do not specifically focus on 'environmental sustainability'. The most challenging of the Rio conclusions is that sustainable development is the overall goal and that economic, environmental, and social development require an integrated approach. It is not sufficient to simply tack on the goal of environmental sustainability to the economic, social, and whatever other goals of development. Without an integrated approach, the extent to which the various goals were implemented would depend on which garnered the most potent backing.

The first Rio principle having declared that human beings are at the focal point of concern about sustainable development, the second principle proceeds thus: 'States have, in accordance with the Charter of the United Nations and the principles of international law, the sovereign right to exploit their own resources pursuant to their own environmental and developmental policies, and the responsibility to ensure that activities within their jurisdiction or control do not cause damage to the environment of other states or of areas beyond the limits of national jurisdiction'. Most of the subsequent principles in the declaration give advice to states on how they ought best to strive for sustainable development and several of these principles call for international cooperation (e.g. principles 5, 7, 9, 12 and 13). But at no point does the declaration state that transnational cooperation would be the only answer to global challenges. Instead, the principle of subsidiarity is emphasised. The argument used to support the principle in this context is that 'Environmental issues are best handled with the participation of all concerned citizens, at the relevant level' (principle 10). It is necessary to differentiate between global and European challenges requiring transnational joint action and those that can be met by national or even local measures.

Land, for example, is a finite resource and the increasing conflicts over land use and the ever-increasing degradation of land resources are a global challenge. *Agenda 21* of the United Nations accordingly regards an integrated approach to the planning and management of land resources as essential and a matter of priority, yet responsibility for promoting the most efficient use of land is not necessarily a task for transnational cooperation. The UN proposes that national governments, in conjunction with appropriate local, national and international institutions and groups, should take action towards outlining a general framework for land use and physical planning while conceding that the majority of local conflicts over land use require a local solution (United Nations 1993, pp 101–104).

Water management needs to be cited as a second example and is discussed in greater depth in this book by Kharitonov, Fischer and Pascariu. As in the case of land use, *Agenda 21* considers the widespread scarcity of freshwater resources and their gradual pollution and destruction as a global challenge calling for integrated approaches to the development, management, and use of water resources. Logically enough, it is stated that 'Water should be managed at the lowest appropriate level, involving district water committees and river catchment authorities', but the UN also calls for joint action at global level: 'At a global level, all nations should aim at the protection of water resources, water quality and aquatic ecosystems...' and later on the UN states, 'An international priority action programme on water for sustainable food production and rural development is relevant...' (United Nations 1993, pp 104–108).

Similar examples could be given regarding human settlements, urban pollution and health, sustainable agriculture, rural development and other aspects of spatial development, but the two examples given may suffice to demonstrate that all around the globe there are challenges requiring cooperation on spatial development to ensure sustainability and that, in many cases, this cooperation needs to be extended to the transnational level.

Values and goals such as sustainable development are fundamental to planning schemes. The first section of the European Regional/Spatial Planning Charter of the Conference of Ministers responsible for Regional Planning within the Council of Europe (CEMAT), drawn up at Torremolinos in 1983, lists the values and overall goals central to the concept of Regional/Spatial Planning: 'Man and his well-being as well as his interaction with the environment are the central concern of regional/spatial planning, its aims being to provide each individual with an environment and quality of life conducive to the development of his personality in surroundings planned on a human scale. Regional/Spatial planning should be democratic, comprehensive, functional and oriented towards the longer term.' This anthropocentric point of departure whereby the main focus is placed on man and his well-being is typical of the discussion on spatial development policy. Irrespective of the fact that 'his interaction with the environment' is also fundamental to spatial development, 'Earth first!' or other ecocentric points of view espoused by environmental groups are not applicable in the European discussion on the development of a transnational spatial development policy. Spatial development has its roots in early urban and regional development schemes of industrial and metropolitan regions which were crying out for a more comprehensive and coordinated development planning system. It was the practical need for a better development of the urban environment in the interests of man that triggered planning efforts and planning policies (regarding Germany see Richter 1994). This approach has remained unchanged simply because Mother Nature has no lobby of her own and there is no empirical way of formulating and verifying what the interests of our natural environment are or could be. Even the sentiment that Nature is in need of sustainable development can be no more than an assumption as to what is in her 'interests'. Man can only underpin the sentiment by either according Nature an ideal value which from a human point of view is considered worth safeguarding or by demonstrating that the sustainable development of our natural environment is a precondition for the sustained development of human society (which is by turn in the interests of man).

In political discussions on a European Spatial Development Policy, sustainability is generally understood as being in the interests of man. A good description of the practical understanding of the relation between the concept of sustainable development and human interests in spatial development can be found in the Leipzig Principles for a European Spatial Development Policy:

> Defined by the Brundtland report of the United Nations and discussed in the Conference in Rio 1992 in Agenda 21, this fundamental concept implies not only economic development which respects the environment, but also balanced spatial development. In this respect it links up with the objective of economic and social cohesion.
>
> Sustainable development, a process of social and economic change, using resources for the benefit of humanity today and in the future, should be considered as a basic principle for the ESDP. The latter can, in return, contribute substantially to the effective application of this new development model.
>
> Sustainability is not a strictly environmental concept, it concerns all human activities. In this approach, economy and ecology become 'two sides of the same coin'. Though conflicts may remain between them for the time being, in the long term environmental quality will increasingly become a prerequisite of economic development. Moreover, it is already an important location factor for investors. Therefore, a sustain-

able spatial structure could be an asset for competitiveness, and stimulate employment. (BMBau 1995, p 43).

3 Sustainable Spatial Development in CEE - the Baltic Example

Though central and eastern Europe (CEE) constitutes a political and economic region, even in the political and economic context the differences between the countries are considerable. As far as spatial development is concerned, central and eastern Europe is more a European space consisting of different and sometimes overlapping European regions such as the Baltic Sea region, the Central European space, the Danubian region and the Black Sea region. Given the differing situations, problems and perspectives of these regions it is not likely that a common CEE spatial development scheme will be agreed in the near future. More realistic and promising is the development of single development schemes for each of these subregions. If they are developed in a similar way and based on the same methods they could subsequently be integrated into a pan-European spatial development scheme.

The European region with the most prominent vision and about five years of practical experience in transnational cooperation in spatial development is the Baltic Sea region (see Zaucha and Fischer in this volume). It is also the European space where cooperation between West and East towards sustainable spatial development is at its most advanced. At their third conference, the ministers for spatial planning and development of the countries around the Baltic Sea approved the report *'Vision and Strategies around the Baltic Sea 2010' (VASAB 2010 or Tallinn Report)* which is intended as a contribution towards a framework for spatial development in the Baltic Sea region and at the same time as an element of the envisaged European spatial development perspective (VASAB 1994; Graute 1996c). In this chapter the focus will be on the values and goals of Baltic Sea cooperation, since the formulation of such values and goals is not only the first step in any process of cooperation but also generally points the way for development.

The *basic values for cooperation in the Baltic Sea region*, aiming at the improvement of the quality of life, are

- development,
- environmental sustainability,
- freedom and
- solidarity.

Beneath these values are ranged *overall goals*. The *VASAB 2010* 'rationale' includes five principal perceptions (VASAB 1994, p 4):

- The Baltic Sea region (BSR) has got to develop a strong identity
- The high degree of spatial cohesion of the past needs to be reactivated and enhanced

- Cooperation helps to maintain peaceful development
- Short-term problems must not prevent the creation of a longer-term perspective
- The region is to become a role model for sustainable development.

Among the basic values cited, sustainability appears only in the context of 'environmental sustainability', while in the list of overall goals 'sustainable development' is mentioned as one of five goals. In the following all five goals will be covered, since the importance of sustainable development as one of five goals can be better understood if it is set against the background of all five.

3.1 Identity

David Kirby (1995, p 439) writes: 'Above all, there is a general determination on both sides of the Baltic to forge a new and distinctive regional identity with which to enter the new millennium.' The Tallinn Report (see VASAB 1994, p 4), says, 'The Baltic Sea Region must develop a strong identity. Because it competes globally with other regions, it needs to develop its own profile and strengths. On this basis, its contribution to a common Europe will gain in value and strength. Investors and other actors need a reliable planning framework. Regional policies can contribute to provide investors and other actors with better information on future regional development.' While Kirby sees a general determination to forge a regional identity, *VASAB 2010* merely posits that this identity must be developed. Undoubtedly there is great interest in cooperation and integration around the Baltic Sea, but the commitment to develop Baltic cooperation and integration does not stand alone but overlaps with the interests of European and global cooperation and integration or, at the very least, competitiveness (Eskelinen, Snickars 1995). Judging by the VASAB quotation, the reason for developing a regional identity appears to emanate more from outside than from within the region itself. The gist is, the Baltic Sea Region needs a strong identity to become globally competitive.

Competition with other nations or regions can of course strengthen regional identity. A basic tenet of sociology is that there cannot be any identity without differentiation between inside and outside. Therefore, to develop a Baltic identity it is necessary to define what constitutes it, where it ends, and what is outside of or actually juxtaposed to it. In the past, geographic proximity and elements of a common history used to be sufficient grounds for a regional identity. But can the same be said to apply in the modern world with its advanced communication systems and mercurial political, economic and social framework? It is certainly not very likely that a strong sense of identification within the Baltic Sea Region will for the foreseeable future extend beyond either the existing Scandinavian or the emerging European identity.

For decades Germany and Denmark have been members of the European Union. Sweden and Finland joined in 1995. Poland and the Baltic States wish to emulate their western neighbours at the earliest possible opportunity. Most countries around the Baltic already have a stronger commitment to the European Union than to the Baltic Sea Region (BSR), notably as regards legal integration. Treaties providing BSR countries with EU associate member status or else full membership in the European Union or European Free Trade Association have had a predominantly economic impact on the countries concerned. In view of the importance which economic relations had in the past for the development of Baltic regional identity, it is even more doubtful that the same identity can develop at a time when European and global economic relations are becoming more and more dominant.

Regions like the Baltic Sea, and indeed probably all the world's regions, are having to deal with a general dilemma. To develop their competitiveness, they need a free market economy and an open society, but to develop their own identity, their own strengths and their own position in the global world, they have got to retain or develop something unique, something different from the rest of the world. For the VASAB process in the Baltic Sea Region, this dilemma constitutes a weakness, while at the same time there is no route in sight that would obviate the need for a balancing act between the two extreme poles (dissociation from the world or total abandonment of awareness of self). *VASAB 2010* reflects this dilemma in the Tallinn Report by striving for both global competitiveness and a regional identity.

The individual inhabitants of the Baltic Sea Region and specially those at the south-eastern coast are already confronted with a second dilemma, viz. the retention and development of several identities. People are being asked to show local, (sub)regional, national, Baltic, European and even global solidarity. If regional identity is so important, then central and eastern Europe is in need of it too, which naturally makes the situation more complicated still. It cannot be expected that every individual will develop a strong identification with all of these levels. To this extent, a strong sense of 'allegiance' towards the BSR, one of the cornerstones of *VASAB 2010*, is either not feasible or will weaken other identifications.

3.2 Spatial Cohesion

This overall goal is fundamental to spatial development in general and is cited as the prime objective of Baltic Sea cooperation: 'Historically, this powerful region was characterised by a high degree of spatial cohesion. This cohesion needs to be enhanced after decades of political division. Social cohesion suffers from unprecedented discrepancies in living standards. Still, however, there is a feeling of belonging together, which is reflected by a remarkable number of cooperation initiatives emerging since the collapse of the iron curtain' (VASAB 1994, p 4). A high degree of spatial cohesion in this region is indeed indispensable. *VASAB 2010* itself gives a description of the situation: 'Interregional discrepancies in living standards: the BSR is an area of great differences in living standards and income. In the group of Nordic countries, differences in average income per capita are small. When crossing the German-Polish border, however, the income level falls sharply, and at the Polish-Belarus border there is a further drop. Belarus achieves only a small fraction of the German level. Between Finland and Russia, a similar discrepancy occurs when crossing only one border line. The Baltic States (Estonia, Latvia, Lithuania) are better off than Belarus or Russia, but a huge difference as compared to Poland or even Nordic and Western countries also exists' (VASAB 1994, p 17).

3.3 Cooperation and Peaceful Development

'Cooperation helps to maintain peaceful development. Cooperation in all possible fields – including regional planning – helps to build a basis of mutual trust and reliability' (VASAB 1994, p 4). Dominant interests in international relations according to Czempiel (1996, p 6) are security, welfare and power. *VASAB 2010* endeavours to support the development of welfare by striving for spatial cohesion, a strong identity, and global competitiveness in the Baltic Sea Region. Geopolitics have often played an important role in international relations, but until recently regional planning with its long-term perspective and comprehensive approach held no importance for the international political agenda. With the increasing international integration of national economies and an increasing number of international environmental problems, the need for coordinated development and regional planning at international level is becoming apparent. The VASAB rationale even considers regional planners, amongst other things, as being co-responsible for maintaining peaceful development. In reality, of course, experienced planners and researchers can only do at international level what they have learned at national/regional level. Is that sufficient, though? *VASAB 2010* says 'no', and makes several proposals in the 'Strategy' section. Included among these proposals are support for international training, advisory assistance, and research networking (compare Vartiainen).

3.4 Longer-term Perspective

'Short-term problems must not prevent the creation of a longer-term perspective. The temptation to do 'first things first' is overwhelming. Sustainable development, though, will not be achieved unless a long-term planning perspective is provided' (VASAB 1994, p 4). This point is listed as a separate pillar in the *VASAB 2010* rationale, but in practice it is already part of the call for sustainable development. It points up a key problem of sustainability, to wit the relationship between economy and ecology. Our modern economy is oriented toward achieving profit within a short-term perspective, while ecological development, in general, is long-term in nature. Economy and ecology often appear to be irreconcilable, but this is only true in as far as our economic thinking remains underdeveloped and hence unable to create a longer-term perspective that would be better able to cope with the conditions of our natural environment.

It takes a lot of time to develop land that is already in use. Accordingly, the fields and disciplines related to this task could not subsist if they were not geared towards the long term. Economists looking for an early return on investment are usually unable to appreciate this perspective. Because of the emphasis on short-term economics in society and politics alike, it is exceedingly difficult to implement long-term projects. This would not be the case were there a comprehensive framework for sustainable development in place. Urban and regional planning for one, and spatial development and research for another, would be among the fields and disciplines which could undergo an elevation of status through the creation of a longer-term perspective for the economy.

The economy would stand to gain, too, of course. As mentioned above, urban and regional planning was only able to evolve because the egotism and short-term interests of increasing numbers of players in the burgeoning cities and metropolitan areas was incapable of unfolding further without the mediation of planners with a longer-term perspective and a more comprehensive approach (Richter 1994). The fact that, in 1994, spatial planners still felt compelled to emphasise the need for a longer-term perspective in a text as fundamental as the *VASAB 2010* rationale shows that far-sightedness can still not be taken as read.

3.5 Sustainable Development

Already the Informal Council of Ministers responsible for spatial development in the EU member states has declared the attributes 'sustainable and balanced' to be a major goal of spatial development. In the Baltic Sea region the Tallinn Report on *VASAB 2010* goes a step further: 'The BSR shall become a masterpiece[1] for sustainable, environmentally sound development. The region has kept valuable natural areas which are now threatened by rapid development. Many cities, however, exhibit environmental degradation due to growing (road) traffic volumes, air and water pollution, required rehabilitation of housing, inappropriate land use, inadequate waste disposal etc. There is also severe environmental damage in other areas requiring remedial action. Hot spots shall be gradually eliminated to the benefit of the water quality of the Baltic Sea' (VASAB 1994, p 4).

It is necessary to emphasise the need for sustainable development. Unfortunately, there are different understandings of what actually constitutes sustainable development. This makes it difficult to establish a consensus for common action. One definition that has found fundamental acceptance is that sustainable development shall be an economic, social, and environmentally sound process. The three respective policy fields are traditionally separate entities. Because of their interdependence in reality, it is obvious that an integrative, all-in approach has to be developed. This is the point at which conflicts usually arise since, in practice, such an approach poses a threat to traditional priorities and privileges. Many environmental studies come to the conclusion that human lifestyles and economic models in the industrialised countries of the world would have to change to make

[1] There is some doubt in our minds as to whether 'masterpiece' can really have been the metaphor intended here. Perhaps 'role model' would have been more apt. [U. G./Trans.]

sustainable development possible. An all-in approach would be an instrument with which to question and eventually halt economic and social development which is not environmentally sound. Because of a lack of experience with this approach, however, insecurities and fears predominate despite widespread appreciation of the theoretical principle.

The handling of the sustainability concept in the *VASAB 2010* Report does not differentiate between the general trend: 'The BSR shall become a masterpiece for sustainable, environmentally sound development.' it states at the outset, but what starts out as a 'masterpiece' has been diluted a few lines later into 'remedial action' and 'hot spots', which are to be gradually eliminated for the improvement of water quality in the Baltic Sea. The adjective 'sustainable' is used several times in the 'Strategies' section; 'sustainable urban environment' (VASAB 1994, p 71) and 'sustainable links between cities' (p 72). Later, in the chapter 'The System: Comprehensive Regional Planning in Function', sustainability is not mentioned at all among the main goals for regional planning. It seems as if sustainability is jettisoned as an overall goal (pp 75–76). This impression gleaned from the text must be erroneous, however, since two years after its publication in their Stockholm Declaration on Sustainable Spatial Development Policy in the Baltic Sea Region of September 1996 the ministers responsible for spatial planning in the Baltic Sea region declare that, 'Subsequently, regional development projects have to be realised in accordance with the criteria of sustainability comprising economic, social and ecological aspects' and that a Regional *Agenda 21* will be outlined for the Baltic Sea Region.

Five years after the UN Conference on Environment and Development and many other meetings on sustainable development there is still no common assessment of the success or failure of the Rio conference. Was it a success or at least partly a success, or was it a failure? Do we now have a better understanding of sustainability and are our efforts towards a sustainable development appropriate? These questions are difficult to answer. Sustainable development is a process with a longer-term gearing and, given the issue, five years is not that long a time. It is thus too early to come to a final conclusion. With respect to the UN recommendations on spatial development (compare 2.3) it can be stated that on the one hand the European discussion on sustainable spatial development has picked up the proposals of the conference and integrated them into political documents such as *VASAB 2010*, *Europe 2000+* and the draft of ESDP. This is an advance. On the other hand the fact that sustainability is often mentioned in these documents in specific contexts such as 'sustainable environmental' (e.g. in *VASAB 2010*) or 'sustained and balanced economic' development (e.g. in *Europe 2000+*) but only occasional as a single all-embracing guiding principle (Leitbild) for future development, as was proposed in Rio, is disturbing. While *VASAB 2010* lists sustainable development as one of several overall goals, the principles for a European Spatial Development Policy adopted in Leipzig at about the same time state sustainable development to be the central aim of spatial development.

Studying the different documents and summarising the discussion process on European spatial development, it becomes obvious that, despite the term 'sustainability' being a key concept for future spatial development, it is applied inconsistently and needs to be tightened up. This process of further definition and clarification will be more difficult than in the case of other terms discussed earlier (cf. 2.1). Intensive application is required in this respect if misunderstandings in future cooperation are to be avoided and circumstances ensured in which it is possible to discuss, develop, and implement sustainable spatial development schemes in a fruitful and ongoing manner. Without an integrated approach as a key distinguishing feature, sustainable spatial development could not be told apart from other approaches and because of this would one day wither away.

References

ARL – Akademie für Raumforschung und Landesplanung (1996) Europäische Raumentwicklungspolitik, Rechtliche Verankerung im Vertrag über die Europäische Union. ARL-Arbeitsmaterial, Nr. 233, Hanover

ARL – Akademie für Raumforschung und Landesplanung (1995) Deutsch-Polnisches Handbuch, Hanover

BMBau – German Federal Ministry for Regional Planning, Construction and Urban Development (1995) Grundlagen einer Europäischen Raumentwicklungspolitik, Principles for a European Spatial Development Policy, Principes pour une politique d'aménagement du territoire européen. Selbstverlag der Bundesforschungsanstalt für Landeskunde und Raumordnung, Bonn

BMBau – German Federal Ministry for Regional Planning, Construction and Urban Development (1994) Newsletter, No. 4. Network of Spatial Research Institutes in Central and Eastern Europe, PLANCO Consulting (ed), Essen, Bonn

BMBau – German Federal Ministry for Regional Planning, Construction and Urban Development (1993) Newsletter, No. 1. Network of Spatial Research Institutes in Central and Eastern Europe, PLANCO Consulting (ed), Essen, Bonn

Council of Europe (1983) European regional/spatial planning Charter. Torremolinos Charter. Strasbourg

Czempiel E (1996) Internationale Beziehungen: Begriff, Gegenstand und Forschungsabsicht. In: Einführung in die Internationale Politik. Manfred Knapp u. Gert Krell (eds), Munich: R. Oldenbourg Verlag, pp 2–26.

EC – European Commission (1994) Europe 2000+ Cooperation for European territorial development. Brussels, Luxembourg

ESDP (1997) European Spatial Development Perspective, First Official Draft. Informal Council of Ministers responsible for Spatial Planning and Development of European Union, Noordwijk, 9–10 June 1997

Eskelinen H, Snickars F (1995) Competitive European Peripheries. Berlin: Springer, Advances in Spatial Science, Berlin, Heidelberg, New York

Forsström B, Vartiainen P, Anderson H (1996) Urban Networking as a Learning Process in the Baltic Sea Region. In: Graute U (ed) (1996) Vision and Strategies Around the Baltic Sea 2010 and their Relevance to Central and Eastern Europe. Dresden: Institut für ökologische Raumentwicklung, IÖR-SCHRIFTEN, vol 17, pp 50–52

Fürst D, Güldenberg E, Müller B (1994) System of Spatial Planning in Germany. Federal Ministry of Regional Planning, Construction and Urban Development (ed) Bonn

Görmar W (1996) German Cooperation in Spatial Planning with Central and Eastern Europe. In: Newsletter, No. 7. Network of Spatial Research Institutes in Central and Eastern Europe, Dresden: Institut für ökologische Raumentwicklung (ed), pp 49–51

Graute U (1996a) The process of *VASAB 2010* in the Baltic Sea Region. European Journal of Regional Development EUREG 4/96: 57–59

Graute U (1996b) The European Dimension in Regional Sciences and the Network of Spatial Research Institutes in Central and Eastern Europe. In: Joris Wotte et al. (eds) Environmental Engineering and Pollution Prevention. NATO ASI Series 2, Environment, vol 18, Kluwer Academic Publishers, Dordrecht, Boston, London, pp 279–292

Graute U (ed) (1996c) Vision and Strategies Around the Baltic Sea 2010 and their Relevance to Central and Eastern Europe. Dresden: Institut für ökologische Raumentwicklung, IÖR-SCHRIFTEN, vol 17

Graute, U (1996d) Umweltpolitik. In: Raumordnung und Raumordnungspolitik. Jenkis H, Munich: R. Oldenbourg Verlag, 1996, pp 287–312

Kirby D (1995) The Baltic World 1772–1993, Europe's Northern Periphery in an Age of Change, London: Longman

Lindström B (1996) Regional Policy and Territorial Supremacy. Nordic Region Building and Institutional Change in the Wake of European Integration. Copenhagen: NordREFO, Report No. 2

von Malchus V (1995) Europäische Raumordnung. In: Handwörterbuch der Raumordnung. ARL (ed), Hanover, pp 261–268

Richter G (1994) Entwicklung der Landesplanung und Raumordnung in Sachsen. Von ihren Anfängen bis zur Auflösung der Länderstruktur. Dresden: Institut für ökologische Raumentwicklung, IÖR-SCHRIFTEN, vol 07, pp 20–42

Savy R et al. (1995) Regions and Territories in Europe, the Region's View of the Territorial Effects of European Policies. Assembly of European Regions (ed), Commission V 'Infrastructure and regional development', Strasbourg

Treuner P (1995) Europäische Regionalpolitik. In: Handwörterbuch der Raumordnung. ARL (ed), Hanover, pp 268–271

United Nations (1993) The Global Partnership for Environment and Development, a Guide to Agenda 21. Post Rio edition, New York

VASAB – VASAB Secretariat (1996) From Vision to Action. Gdansk

VASAB – VASAB Secretariat (1995) Statements Relating to the Report

VASAB – Group of Focal Points (1994) Vision and Strategies around the Baltic Sea 2010. Towards a Framework for Spatial Development in the Baltic Sea Region. Tallinn Report, prepared for the Third Conference of Ministers for Spatial Planning and Development. December 7–8, 1994. A summary of the Tallinn Report, compiled by Bo Löwendahl (1995), was published in German and English in Newsletter 5 of the Network of Spatial Research Institutes in Central and Eastern Europe. Institut für ökologische Raumentwicklung, Graute, U. (ed), Dresden, pp 11–32

Part I

Towards a European Spatial Development Policy

Sustainable Development for Greater Europe

Tarcisio Bassi
Council of Europe, Directorate of Environment and Local Authorities
Strasbourg, France

The Council of Europe has been active in the area of regional/spatial planning since 1970, when the first meeting of the European Conference of Ministers responsible for Regional Planning (CEMAT) was held in Bonn.

The Ministers have subsequently met again on several occasions. One of the most important decisions was taken at Torremolinos in 1983 when the European Regional/Spatial Planning Charter was adopted.

The Charter clearly sets out the basis for regional/spatial planning: it should be democratic, comprehensive, functional and long-term oriented:

- democratic, so as to ensure the participation of the people concerned and their political representatives;
- comprehensive, to coordinate all sectoral policies;
- functional, to take account of regional differences and the characteristics of the different countries;
- long-term oriented, to take into consideration the long-term trends of economic, ecological, social and cultural phenomena.

At their last meeting in Oslo in 1994, the European Conference of Ministers responsible for Regional Planning within the Council of Europe adopted an action programme which focused on long-term analysis as the indispensable means for enabling public authorities to take coherent decisions and devise strategies for the promotion of sustainable and balanced development.

At the next CEMAT session, to be held in Cyprus in 1997, the Ministers will exchange views on "Sustainable regional/spatial planning in Europe and the protection of water resources".

The political upheavals which occurred at the beginning of the decade, with a speed and to an extent unprecedented in the history of modern Europe, created an unforeseen need for integration between the countries of central and eastern Europe on the one hand and western Europe on the other.

At the same time Europe as a whole was, and continues to be, confronted with major changes affecting the very future of regional planning, for example:

- the growth in world population;
- the demographic stagnation and ageing of the European population;
- the growth in trade and its impact on transport;
- the delocalisation of enterprises;

- the development and impact of European sectoral policies, in particular regional policy and the common agricultural policy;
- the world-wide emergence of lifestyles based on over-consumption of natural resources;
- dissatisfaction with existing social systems stemming from the current economic problems and the incessant rise in unemployment.

The contemporary world is changing very rapidly. Problems are becoming more complex, technical and interrelated. As past experience can throw little light on future development, it is difficult to assess the medium and long-term impact of current lifestyles on the future. Serious thought needs to be given to the interests of future generations. Social justice should apply not only to the present generation but also to those of the future.

Future development must be sustainable and meet the needs of the present without undermining the ability of future generations to satisfy their own needs. All too often, however, short-term concerns take precedence over the needs of the future.

There is therefore a greater need than ever for regional/spatial planning involving the search for strategies that can achieve balanced long-term development in the economic, social, ecological spheres etc. and as a means of coordinating the various sectoral measures and policies proposed by those working at a regional, national and European level.

The growing internationalisation not only of industrial and agricultural production but also of all the other factors which shape our lives and determine the life of tomorrow's citizens means that regional planning problems cannot be dealt with at national or regional level alone.

Economic development over the last decades has been unevenly distributed. It has been incapable of significantly reducing inequalities either between European states or between regions within the same state. We need to consider the possibility of establishing greater solidarity – without which there can be no lasting social peace – within Europe, and the ways and means of achieving it.

Greater Europe will be unable to maintain its present position in the world and in the economic, cultural and political arenas unless it can find solutions to the major challenges facing it.

The trend in this nascent and inchoate Greater Europe is to move away from the previous North/South spatial organisation of Europe towards a West/East orientation. This reorientation will have serious repercussions for economics, transport and population flows.

This spatial reorientation should not make us lose sight of the special problems facing the countries of southern Europe, which have large young populations who also aspire to a higher standard of living. We must take account of the problems of all European countries and have the courage to establish priorities as it is impossible to do everything immediately or at the same time.

CEMAT believes that a high priority should be given to analysing the effects on European regional planning of the social and political changes in eastern and central Europe.

It is important to recognise that, given the serious disparities which exist between the countries of Greater Europe, the political changes in central and eastern Europe which have paved the way for the Greater Europe of tomorrow not only offer new opportunities but also pose many problems.

Not all countries of Greater Europe will benefit to the same extent from the gradual rapprochement and growing economic interdependence between the countries of central and eastern Europe and the industrial countries of the West. Change will be a slow and long-term process. During the period of transition and before a new stable equilibrium is established, it will be necessary to monitor carefully the development of these changes and try to evaluate their impact on:

- areas where the consequences will be most significant;
- countries and regions which in the immediate future and in the longer term will be adversely affected by these changes;
- on border regions;
- forms of cooperation which can bring about closer cohesion between peoples and territories and at the same time reduce the disparities between East and West;
- measures to strengthen the current democratic process and to promote greater political stability.

One key area where the Council of Europe can and must be active in the future lies in the elaboration of medium-term strategies for achieving balanced development in Europe.

Economic development, it needs re-emphasizing, has been unevenly distributed over the past decades. However the point at issue is whether it is possible to reconcile sustainable development and balanced regional development and at the same time meet the ever more demanding challenges of international competition.

Should priority be given to effectiveness and therefore to specialisation and the concentration of economic activity in the most competitive regions, with a consequent shift in population towards the most economically successful regions, at the expense of balanced development; or should we give priority to regional balance and promote the establishment of new enterprises in or the transfer of economic activity to the least developed regions as a way of distributing development more equitably throughout Europe as a whole, with the consequent need to assist and subsidise enterprises whose products are unable to compete nationally and internationally? Experience shows that it is very often difficult to sustain this type of development once the subsidies are removed.

Should we only strengthen and develop those economic activities where Europe excels and is most competitive and gradually phase out economic activities based on the extensive use of an unskilled workforce or on the large-scale exploitation of natural resources? What would be the social costs of this choice and how could they be met?

The new Greater Europe will undoubtedly call into question certain policies to help the less favoured European regions and the assistance given to certain industrial and agricultural activities which are in decline and which can only survive through reconversion. A large number of regions will be affected by these changes. New disparities will be created between prosperous and dynamic towns which have been able to coordinate their resources and network their activities, and other towns with fewer resources which lack foresight and try unsuccessfully to continue as in the past. Changes in the structure of territorial Europe will increase regional disparities and imbalances between more competitive and less competitive areas. If steps are not taken to reduce these disparities, European economic and social cohesion may be jeopardised and population shifts towards the major cities and economic growth centres will continue unabated with adverse consequences both for the depopulated areas and for those where population concentration creates a wide range of problems which are increasingly difficult to solve.

It is important from a political point of view to ensure that the development of individual European regions or population groups less able and less well equipped to adjust to the changing modern world does not lag behind unduly. Peace itself may well depend on whether proposals regarding ways and means of keeping the disparities in economic development in Greater Europe within acceptable limits succeed or not. The economic integration of a large number of European countries and even wider political cooperation will lead to increasing demands from the different regions of Europe for similar living conditions and comparable economic growth.

At the same time, Europe must not spread its efforts too thinly. Although it will clearly have to take short-term defensive measures to deal with crisis situations, it will simultaneously have to adopt long-term development measures to safeguard its future.

Regional planning policy must be neither entirely liberal nor entirely interventionist. For liberalism, the only imbalances which cause concern are those which prevent the free play of world forces. According to the interventionist point of view, space should be organised to promote more balanced liberal economic development.

Regional development policies in Europe will probably have to aim at qualitative rather than quantitative growth and at establishing links between towns and villages to prevent the development of inhuman conglomerations on the one hand and depopulated areas on the other.

CEMAT seeks to provide policy-makers with coherent strategic choices for the future so that appropriate solutions can be found to the regional planning problems referred to above, which concern both the older Council of Europe member states and the new ones. A sustainable and satisfactory solution will only be reached through international consensus.

For this reason, CEMAT is currently preparing a document setting out the guiding principles of sustainable overall planning in Greater Europe for the next century. The European Union is also working on the same topic as it affects the 15 member states of the Union. CEMAT's aim is to draw up a non-binding text, more detailed and precise than the European Charter on Regional/Spatial Planning adopted thirteen years ago, which will provide a coordinated framework of long-term guidelines for all the states of Greater Europe. This text will not be easy to prepare because of the need to reconcile not only the points of view of the member states of the Council of Europe but also those of the regions in whom responsibility for regional policy is frequently vested. It will also be necessary to change priorities and lend greater weight to medium-term solutions than to current problems. In short this means putting the living standards of Europeans on hold or even lowering them slightly in some cases.

Work on the document has already begun and discussions have taken place in the committee responsible for the organisation of CEMAT. A number of ideas have already been approved. The document should:

- take account of the basic principles approved by the 15 member states of the European Union in drawing up the European Spatial Development Perspective (ESDP);
- be based on the principles of the Torremolinos Charter, while at the same time representing a more practice-oriented step forward;
- not be binding on the states;
- take account of the principle of sustainable development, of the rights of future generations and the effect of the political enlargement of Europe;
- promote equitable development throughout the whole of Europe.

It is intended to finalise the document for the 12th session of CEMAT and to submit it for adoption at the meeting in Berlin in the year 2000.

However, to achieve tangible results it will not be enough simply for specialists to work together and to prepare yet another document. The general public must be informed of, understand, and be convinced of the need for the proposed changes, which as well as impacting on the lives of tomorrow's Europeans, are also both necessary and indispensable if a viable quality of life is to be forthcoming for the six billion individuals who will inhabit the planet as a whole at the beginning of the next century. Young people who, more than us, will have to bear the consequences of the success or failure of decisions taken (or not taken) today, should be given a greater voice in the major strategic choices that will determine our future. Would it be undemocratic in these cases to fix different age limits for participation in referenda from those which already exist? We should not forget that the world population has more than doubled in half a century. It is clearly impossible to extend our consumer habits to the planet as a whole. However, since the acceptable limits differ from country to country, individual countries often act without taking account of global limits.

Despite the serious and difficult problems which need to be solved, many of which are the results of continually postponing unpopular decisions which should have been taken earlier, it is important not to lose confidence in the ability of human beings to survive and adapt to change, this having always been a feature of life on our planet. In any event, we must try, with all due modesty, to predict these changes and guide them in the direction of harmonious, balanced and democratic solutions.

Sustainable Development in Europe – A Challenge for Transnational Spatial Planning Policies

Welf Selke
German Federal Ministry for Regional Planning,
Construction and Urban Development
Bonn, Germany

1 Competitive Imbalances between Cities and Regions on the Territory of Europe

The general political conditions for the development of cities and regions in Europe have changed radically. Europe is undergoing structural change with far-reaching consequences. With globalisation proceeding apace, the development of innovative areas and new technologies has a direct impact on regional economic structures and labour markets. Since the Iron Curtain was dismantled the national economies of all countries in Europe have been converging to form a pan-European market. Association agreements concluded by the European Union over the past few years with the countries of central and eastern Europe (CEE) are, together with partnership agreements concluded with the Russian Federation and Ukraine, an important step towards integrating these countries into a Common European Economic Area.

The enlargement of the Council of Europe from 24 states (1990) to 40 states (1996) signals a willingness amongst European States to cooperate in a great number of political spheres on the common basis of the principles of democracy and human rights. This process has also consequences for the development of local and regional democracies in the new member states.

With the opening up of what had been politically sealed economies, the prospects for development have now become more difficult for a large number of cities and regions in Europe. In future, cities and regions will have to compete for investment in a more uniform economic area. Their locational advantages and disadvantages within the European territory will have a significant bearing on the potential prospects of their inhabitants.

The transition of the former communist States in central and eastern Europe (CEE) and in the Commonwealth of Independent States (CIS) from centrally-planned to market economies has naturally had significant consequences in terms of regional competitiveness and development. The collapse of the central planning system has had a sudden and negative impact on options for future development in the regions concerned. Many towns and regions in the former communist bloc depended on a single State enterprise or a small number of such enterprises, which had no economic grounding but received government support for reasons of a superimposed political nature. The introduction of market structures means that the economic advantages and disadvantages of the various towns and regions as locations have now come fully into play again. As the new markets have opened up, foreign investment has mainly flowed into the capital cities of the old and new States of central and eastern Europe, or into the odd economically attractive region, thus contributing to the confinement of economic growth to regional pockets.

The widening gap in terms of social and economic development between a small number of privileged regions and far greater areas of the countries concerned in which the impact of economic transformation has been negligible carries with it the risk of political tension in the medium term. That is why there is a need for innovative regional development programmes to ensure that as many towns and regions as possible can benefit from the positive effects of the new economic and social order. This is the task of spatial planning policy, which in many transition States is evolving into a new area of politics. However paradoxical it may seem following the demise of central planning, the social market economy also needs a democratically based system of planning where regional development is concerned.

2 The Transnational Dimension of Spatial Planning Policies in Europe

In the case of much national investment, be it private or public, it is not sufficient to view or evaluate its regional impact purely from a national territory point of view. Many funds and sectoral policies apply to spatial development at the cross-border, transnational or European level. National spatial planning policy has to take account of these impacts. That is the reason why national spatial planning authorities in several major European regions have drawn up transnational strategies for spatial development.

The European Commission and the 15 Member States of the European Union have been working since 1994 on the preparation of a European Spatial Development Perspective (ESDP), which is intended to provide a guiding framework facilitating attainment of the objectives of economic and social cohesion agreed in the Treaty of Maastricht (1992). In the view of both the Member States and the Commission, achievement of what is known as the cohesion objective requires sectoral as well as structural policies.

The European Spatial Development Perspective (ESDP) discussed by the EU Member States at the informal meeting of Ministers responsible for Spatial Planning Policy at Noordwijk (Netherlands) in June 1997 provides the European Union with a very robust tool for work in this field (see: van Ginneken). This document is the culmination of preparatory work carried out during the past six Presidencies of the EU. The principles, operational objectives, and three spheres of activities laid down at Leipzig (1994) form the basis of the ESDP.

During the German Presidency of the Council, the then 12 Member States of the EU and the four countries wishing to join, Austria, Finland, Norway and Sweden, reached agreement on the "Principles for a European Spatial Development Policy" at their meeting of the informal Council of Ministers held in Leipzig on 21–22 September 1994. This transnational spatial development perspective declares sustainable development to be the strategy with which to fulfil the principal objective of attaining economic and social cohesion in the European Union. Under this approach, economy and ecology are two sides of the same coin and hence form the dual core of a well-targeted regional development control mechanism. True, there may at present be conflicts between the two elements but, in the long term, the quality of the environment will increasingly become a precondition for economic development. Moreover, it is already an important locational factor for investment. A sustainable spatial structure could therefore promote competitiveness and stimulate employment.

Economic and environmental development in the EU must be based on a more balanced spatial structure. As stated in the basic document adopted in Leipzig spatial planning policies must combine the following objectives:

- the development objective: to restructure and strengthen the structurally weaker areas of Europe, which are of particular importance for the territorial cohesion of the European Union;
- the equalisation objective: to bring into line the living and working conditions of the population beyond national boundaries between areas of different levels of development;
- the conservation objective: to preserve, re-establish and connect within a balanced and transnational network a coherent system of open spaces, with their respective functions.

In order to achieve these operational objectives of spatial development, the Leipzig paper recommends that the European Spatial Development Perspective should promote the following spatial planning policy initiatives at European level:

- A polycentric and well-balanced system of city regions which seeks to avoid excessive concentration in a small number of large centres and the marginalisation of peripheral areas;
- A network of environmentally sound and efficient infrastructures that reinforces the territorial cohesion of the Community area;
- A European system of open spaces improving the protection of natural resources with graduated protection areas corresponding to its various functions.

The Leipzig paper recommends six main planks of spatial development policy in the European context (table 1). Among these political measures it was agreed to work together for the establishment of an "observatory" in which the relevant institutions of the Member States would cooperate on research topics and analyses designed to back up major decisions on Europe's future spatial structure and development. In 1996, as a means of promoting spatial-planning pilot projects in the European context, the European Commission launched a new programme dubbed INTERREG II C (see chapter 4).

In Noordwijk the Spatial Planning Ministers and the Commissioner for Regional Policy and Cohesion endorsed the first draft of the European Spatial Development Perspective, which will now form the subject-matter of a broad political debate to take place:

- within Member States, both with the relevant sector administrations and the regional and local authorities;
- between Member States at transnational level, thereby contributing to a better integration of action based on the ESDP and on INTERREG II C;
- between the European Commission and other European institutions;
- with neighbouring countries outside the European Union, in particular the associated countries (in collaboration with the Council of Europe).

Those involved underlined the importance of getting this political debate going without delay, with the aim of drawing up the format ESDP document by the end of the United Kingdom Presidency in 1998.

Similar close cooperation between spatial planning authorities is likewise already taking place outside the EU, in the Alps, for example, within the framework of the Alpconvention.

Table 1. Principles for a European Spatial Development Policy
Leipzig, 21–22 September 1994

Operational objectives of spatial planning	Development	Equalisation	Conservation
	OBJECTIVE		

Strategy	Sustainable development

Spheres of activity covered by spatial planning policies	Networks of Urban systems	Parity of Access to Infrastructure and Knowledge	Wise Management and Sustainable Development of Europe´s Natural and Cultural Heritage

Political measures (implementation of spatial planning policies):

1. Strengthening spatial planning in the EU Member States.
2. Promoting cooperation between the Member States.
3. Seeking cooperation with non-Member States.
4. Improving the coordination of Community policies (specific role of the Commission).
5. Creating a spatial development observatory.
6. Promoting spatial planning pilot projects in European regions.

3 The Baltic Sea Region: A Model for Pan-European Spatial Planning Cooperation with CEE States

It is not possible to apply the very far-reaching aims and objectives for spatial development of the 15 Member States of the EU to the whole of Europe. Funds are lacking for a spatial development policy of this kind and there are no pan-European policy areas within the framework of, for example, the Council of Europe with corresponding decision-making powers to influence spatial structures. The main aim of a European transnational spatial development strategy must be to clarify the role of spatial development policy in the countries of central and eastern Europe in the economic and social development of a space. The exchange of information and the transfer of modern know-how in the field of spatial planning policy must also be promoted.

The Baltic Sea Region can serve as a good model for cooperation among the countries of Europe notwithstanding the former Iron Curtain. It is here that the greatest progress in transnational cooperation between EU Member States and transition countries has so far been achieved (see: Zaucha and Fischer). The third conference of Ministers for Spatial Planning and Development held at Tallinn in 1994 adopted a transnational framework for spatial development known as "Vision and Strategies around the Baltic Sea 2010" (VASAB 2010).

The VASAB 2010 Perspective for the Baltic region is rooted in the goal of achieving what is now referred to as sustainable development. It will be important to make sure "sustainability" is not seen by any of the states around the Baltic as a form of "de luxe planning" undertaken by the rich Member States of the EU but rather as something serving to significantly enhance the development prospects of the cities and regions of Europe in the light of global competition. In addition, the following points are addressed as separate basic values:

Freedom: this means here the opportunity for comprehensive democratically legitimised participation in the processes of regional and local planning and development.

Subsidiarity: The guarantee of subsidiarity is an important instrument for facilitating participation. The principle of subsidiarity contributes to improving and increasing the quality and the political acceptance of decisions.

Solidarity: solidarity is a feature of a welfare society that enables the advantages of development to be shared by the entire territory.

Spatial planning policy in the Baltic Sea Region as a whole will promote these three basic values. The spatial planning policy concept of VASAB 2010 sets out the main areas of application of spatial planning in a vein similar to that of the basic document adopted in Leipzig:

- strengthening the network of urban entities of European, national and regional significance (the "Pearls");
- linking these urban regions in an effective and sustainable manner with the aid of transport and energy networks (the "Strings");

- assisting areas with special problems and development potentialities such as islands, coastal regions, natural landscapes, border areas (the "Patches").

Comparing the two transnational spatial planning development strategies of the EU and the Baltic Sea Region it becomes clear how much they have in common as far as the objectives and areas of application of spatial planning are concerned. Yet there is one marked difference: spatial planning policy in the EU aims to a large extent at achieving spatial convergence by basing Community policies to a greater extent on spatial planning perspectives, whereas spatial planning policy in the Baltic Sea Region seeks to promote a democratically legitimised and decentrally organised spatial development process.

In the Baltic Sea Region efforts are now being made to translate joint spatial development objectives and jointly defined areas of application into concrete transnational projects.

The report "From Vision to Action", which was adopted at the last (4th) Conference of Ministers responsible for Spatial Planning held at Stockholm in 1996, lists the main forms of spatial development project targeted for implementation over the next few years on the basis of close cooperation among national authorities. There are six project areas of particular significance to spatial planning policy:

- pilot projects to foster urban cooperation;
- use of transport corridors to improve sustainable regional development;
- cooperation on marine transport;
- sustainable tourism (eco-tourism);
- transfrontier cooperation in the fields of spatial planning and urban development;
- exchange of information through the holding of seminars, preparation of joint handbooks, etc.

In carrying out these projects, close cooperation between the spatial planning authorities of the Member States of the EU and the fledgeling democracies is taking place. The aim is to integrate the countries in transition into the process of European cooperation on spatial development. To illustrate this, it would be helpful to mention two projects Germany is involved in.

Cooperation on spatial planning policy in the German-Polish border region has led in a first development to the elaboration of joint "spatial planning perspectives for the area along the German-Polish border". These have been enshrined in government resolutions in both countries. An initial field project – more specifically one concerned with developing tourism in the German-Polish Baltic Sea Region – was successfully launched in 1996. The project provides Poland and Germany (Mecklenburg-Western Pomerania in this instance) with a valuable basis for decision-making and planning principles for the development, expansion and suitable integration from a locational point of view of a network of marinas. This is now to be extended to other coastal areas in order to promote an environmentally sound form of tourism based on sailing.

Another priority for spatial planning cooperation in the Baltic Sea Region is the use of transport corridors to improve sustainable regional development. The second pan-European Conference of Ministers responsible for Transport, held from 14–16 March 1994 on Crete, established nine priority corridors in the countries of central and eastern Europe. These corridors constitute extensions of the large transport axes of the western European transport network to central and eastern Europe. Five of these corridors affect the Baltic region. It is easy to predict that these will form major focal points for infrastructure investment activity in the coming decades.

These measures should not be judged in transport policy terms alone; rather they provide an opportunity for generating regional development. Thus, a regional development concept is being prepared for a section of Trans-European Corridor II (Berlin–Warsaw–Minsk–Moscow), which traverses territory of the Republic of Belarus, as a means of promoting decentralised economic development along the corridor. In this way, the general conditions for investment are also being improved for the difficult economic environment outside the region of the capital Minsk (compare Ivlichev, Khazdan).

There is a growing demand in the Baltic Sea area for the various instruments and methodologies of spatial planning policy. Since a regional impact assessment and/or the optimisation of investments in transport corridors require the cooperation and active participation of all administrative levels and of various private and public corporations, they represent a useful means of making market-oriented spatial planning policy more accessible to the countries in transition. This type of project-oriented cooperation based on the principle "learning by doing" has proved to be extremely fruitful in the Baltic Sea Region and is influencing the success of spatial development strategies in this area.

4 Towards a Pan-European Sustainable Spatial Development Strategy

The examples of the EU and the Baltic Sea Region demonstrate that, as national economies converge to form a pan-European market, it is necessary to support this process by pursuing a pan-European spatial development strategy. The Council of Europe (European Conference of Ministers responsible for Regional Planning/CEMAT) is engaged in efforts to harmonise and coordinate spatial planning policies. The European Regional/Spatial Planning Charter of 1983 provided a first set of common principles, but countries of the communist bloc were excluded at that time. The fact that almost all European countries are now members of the Council of Europe provides an opportunity to take the 1983 charter further and transform its ideas into practical action (see: Bassi and von Malchus).

Germany will host the twelfth European Conference of Ministers responsible for Regional Planning in the year 2000 in the context of the World Exhibition (EXPO 2000). The focus of this conference will be on the consideration of guidelines for a pan-European spatial development strategy. The main aim of this strategy should be to strengthen spatial cohesion within the framework of the sustainable development of European cities and regions in an "expanding Europe".

Help is provided here by the European Union. The new Community Initiative entitled INTERREG "Spatial Planning" (II C) provides spatial planning authorities in the Member States and those in third-party countries with the opportunity to cooperate on joint spatial development projects. Among the existing instruments agreed at European Community level, INTERREG II C is the one that most explicitly supports operational spatial planning programmes at a transnational level. The INTERREG II C programme can be seen as signalling that the link between spatial planning policy and the EU's regional policy, and thereby between economic growth and transnational spatial planning, will be strengthened in the coming years.

Furthermore, Article 10 of the European Regional Development Funds (ERDF) provides for pilot initiatives being carried out in that field. While the EU support programmes for the transition States, i.e. PHARE, TACIS, and MEDA (for the Mediterranean non-Member States), do not make explicit reference to spatial planning, they can nevertheless serve as an important means of supplementing INTERREG programmes in the area beyond the EU having regard, amongst other things, to accessibility and sustainability.

There are 14 Operational Programmes to be implemented under INTEREG II C Seven involve general cooperation in the field of spatial planning and development policy (see table 2), three address themselves to flood prevention, and there are four national programmes for the prevention of drought through prudent water management. Programmes promoting transnational cooperation in the sphere of spatial planning have a total budget of 224 million ECU and will embrace almost the whole of geographical Europe, since two of the cooperation areas will include the transition countries in central and eastern Europe (see table 2 and figure 1).

The European Commission is currently deliberating on how to enable joint spatial development activities in the Baltic Sea Region and in the Central, Adriatic, Danubian and Southeastern European Space (CADSES) to receive funding from both the INTERREG initiative and the PHARE and TACIS support programmes. Only by combining these various European Commission support mechanisms will it be possible for transnational spatial development projects jointly initiated by EU Member States and transition countries and aimed at promoting spatial cohesion to be carried out effectively.

Table 2. INTERREG II C Trans-European Cooperation Areas

Programmes/Areas	Participation States	Total budget - Mio. ECU -
ATLANTIC SPACE	France, Ireland, Portugal, Spain, United Kingdom,	22,2
BALTIC SEA	Germany, Denmark, Sweden, Finland and Baltic Sea States	47,6
CADSES: CENTRAL ADRIATIC, DANUBIAN AND SOUTH-EASTERN EUROPEAN SPACE	Austria, Germany, Greece, Italy and CEE-States	39,6
MEDITERRANEAN and LATIN ALPS	France, Italy, Spain,	22,8
NORTH SEA	Denmark, Germany, Netherlands, Norway, Sweden and United Kingdom	29,7
NORTH-WESTERN METROPOLITAN AREA	Belgium, France, Germany, Ireland, Luxembourg, Netherlands, United Kingdom,	53,1
SOUTH-WESTERN EUROPE	France, Portugal, Spain	9,0

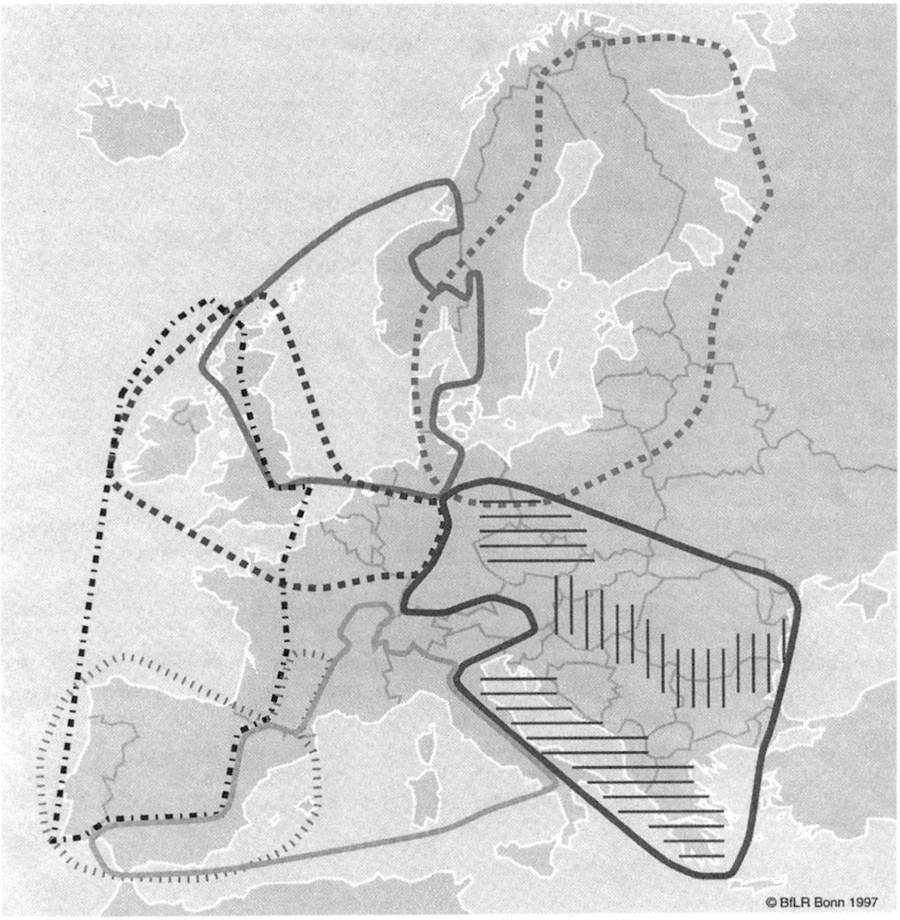

INTERREG II C

- Baltic Sea Region (BSR)
- North Sea Region
- North-Western Metropolitan Area
- Central Adriatic, Danubian and South-Eastern European Space (CADSES)
- Western Mediterranean and Latin Alps
- Atlantic Space
- South-Western European Space

Figure 1. INTERREG II C Transnational Cooperation Areas
Present state of discussions between the EU member states (July 1997)

In tandem with political discussion of the ESDP and of the envisaged pan-European spatial development strategy, implementation of INTERREG II C trans-national programmes may in the years ahead well prove to be the decisive driving force along the road towards a more balanced development of the regions in a continent that is busy evolving into an organic whole. These, along with spatial planning policies at national level, are important pre-conditions for sustainable development. To the extent that their respective legal and political frameworks allow this, the European Union and the Council of Europe will do their utmost to complement each other in pursuing such an objective.

References

Council of Europe (1996) Colloquium on Regional Planning at the Dawn of the Year 2000, Strasbourg 29–30 October 1996, AS/CEMAT/Coll. Str.

European Commission (1994) Europe 2000 + Cooperation for European territorial development, Luxembourg

European Spatial Development Perspective (ESDP) – first official draft, presented at the informal meeting of Ministers responsible for Spatial Planning of the Member States of the European Union, Noordwijk, June 1997

Federal Ministry of Regional Planning, Construction and Urban Development (1994) Principles for a European Spatial Development Policy, Bonn

Krautzberger M, Selke W (1996) Perspektiven der bundesstaatlichen Raumplanungspolitik in der Europäischen Union – Das Beispiel Bundesrepublik Deutschland; Schriftenreihe Stadt und Region, Bd 5 (Hrsg Institut zur Erforschung von Methoden und Auswirkungen der Raumplanung der Ludwig Boltzmann Gesellschaft), Wien

Secratariat of Vision and Strategies around the Baltic Sea 2010 (1997) From Vision to Action – results of the 4th Conference of Ministers for Spatial Planning and Development, Gdansk

Towards a European Spatial Development Perspective

Karolijn M. van Ginneken
Ministry of Housing, Spatial Planning and the Environment,
National Spatial Planning Agency
The Hague, The Netherlands

1 Introduction

The ministers of the European Union responsible for spatial planning have been convening approximately twice a year since 1989. These meetings reflect a need for communication between the Member States of the European Union over issues relating to spatial planning beyond one's own national boundaries.

In June 1997, under Dutch Presidency, an informal conference of the EU ministers responsible for spatial planning was held in Noordwijk at which the first draft of the European Spatial Development Perspective (ESDP) was presented. This document forms the basis for a common European agenda on spatial planning. Consensus between the Member States is essential when developing a plan such as the ESDP. Building this consensus is a process that is time-consuming. The first draft of the ESDP can be seen as a step forward in this process. As agreed at Noordwijk, the next step will involve presentation of the first official ESDP during the British EU Presidency in June 1998.

Hitherto, spatial planning has not been a formal area of policy on the European agenda. As it falls under the subsidiarity principle, spatial planning is not a Community task and as such is not mentioned in the Maastricht Treaty signed by all Member States in 1991. Nevertheless work has been proceeding at a European level.

The first Section of this chapter sets out the motivating forces that led to the ESDP being drawn up. Section 2 focuses on the status of the ESDP. Questions answered here are: how far does the ESDP go, and what is the strategy behind it. After reviewing the evolution of the ESDP and the steps taken towards bringing the document about in the third Section, the contents of the ESDP are dealt with in Section 4. Territorial analysis represents an important aspect of the ESDP that therefore deserves attention in Section 5 of this chapter. As the European Union is only part of the continent of Europe, the link between the ESDP and central and eastern Europe is made in Section 6. Finally, the future of the document and the implementation of its contents are discussed in Section 7.

2 Origins

As a consequence of the unification of Europe and the globalisation of the economy, spatial issues are becoming increasingly international in their orientation. Spatially relevant decisions and developments in one country are having an increasing impact on neighbouring countries. Moreover, the growing territorial impact of Community policy on the spatial organisation of Member States necessitates cooperation directed at establishing a common spatial planning policy.

"A growing interdependancy between the different parts of the European territory accompanies this trend. Three different forms of interdependancy can be defined: (1) between territorial entities; (2) between the various sectoral policies which have an impact on the territory (3) between the various levels of government which share responsibility for a particular territorial unit" (ESDP, 1997).

Given these factors, the needs can be defined that have been identified by the Member States of the EU. The Member States feel that, to be able to develop an adequate spatial planning policy, they should gain more insight into cross-border and transnational processes. Decisions on land use, of an infrastructural nature for instance, have an impact that transcends national borders, making transnational cooperation imperative. An awareness of the need for greater control over policy decisions made outside national boundaries as well as over policy decisions made at supra-national level has arisen. In addition, the Member States have realised that these policy decisions should play a role in new spatial policy development at all levels of government.

Three fundamental goals, set to face the spatial developments in Europe, are pursued in the ESDP:

- Economic and social cohesion
- Sustainable development
- Balanced competitivesness of the European territory

When looking to the future, the long term trend of economic, social and territorial development of the EU will be affected by three further factors: "(1) the consequences of the gradual economic integration of Europe, and of a more intensive political cooperation; (2) the growing role of the local and regional authorities as well as their traditional role on spatial development; (3) the likely enlargement of the Union. The changes mentioned above sometimes create the risk of increasing territorial imbalances in Europe but frequently they also offer fresh opportunities for progress. In either case, they call for concerted responses from the Union and point to the benefit of further developing the territorial approach at a European level" (ESDP, 1997).

The authorities with responsibility for territorial development in the Member States and at Community level face new challenges in the wake of these changes. By reaching an agreement on the development of a spatial perspective of Europe, these authorities have expressed their understanding of the new challenges and indicated that an approach which transcends national boundaries is necessary.

3 Status of the ESDP

"The ESDP goes further than just laying down objectives from a territorial point of view. It also deals with policy issues. It is a document which aims both to raise the awareness of the various partners involved (Commission, European Parliament, the various Council bodies, the Member States, regions and other potentially interested parties) and to promote the organisation among them of a wide-ranging discussion on the long-term guidelines for European territorial development" (ESDP, 1997).

The overall territory of the Union is large and diverse. The ESDP therefore includes an analysis of the entire territory and its features. Policy strategies set out in the document should take these features into account and encourage coherence and complementarity between Member States. An integrated, multi-sectoral and indicative strategy is what is needed. Furthermore, the document should be concerted, selective, progressive, on-going, flexible and transparent. Another important feature of the ESDP is the fact that it should be indicative, and not prescriptive. "Nevertheless, it is expected to lead to tangible results for the European territory and provide a stimulus for action" (ESDP, 1997).

The ESDP is not geared to creating new responsibilities and new policy fields but rather to coordinating existing policies and increasing their relevance by paying attention to the territorial dimension.

Territorial matters are dealt with in three spheres of activity defined at the ministerial meeting in Leipzig. Through these spheres of activity, or "guiding principles", a structural approach to integration and cooperation can be striven for. The following guiding principles have been identified:

- a balanced and polycentric urban system
- parity of access to infrastructure and knowledge
- prudent management and development of the natural and cultural heritage.

4 History

In 1991 the "Informal conference of the European Ministers responsible for Spatial Planning and Regional Development" decided to set up a Committee for Spatial Development (CSD). The CSD is contained within DG XVI (regional policy) of the European Commission. Its aim is first of all to generate an exchange of information. Secondly the aim of the CSD is to address itself to the procedures involved in communications between Member States and the Commission regarding issues relating to the development of the territory of the Union. The tasks of the CSD are to conduct research on issues pertaining to spatial development at Community level and to formulate a perspective on the future development of the Union's territory.

In line with the remit of the CSD, ministers confirmed the necessity to produce a European Spatial Development Perspective (ESDP) at the informal conference in Liège in 1993. It was explicitly stated that the drawing up of the ESDP is the combined responsibility of the 15 Member States and the Commission. From that time onwards, work on the ESDP has been continued by the respective Union Presidencies. In the ESDP process, the following acts were instrumental in developing ideas on the status and content of the document:

- During the ministerial gathering on Corfu in June 1994, the structure and index of the ESDP were decided upon.
- The policy guidelines of the ESDP were laid down under the German Presidency when a document entitled "Principles for a European Spatial Development Policy" was presented to the convention of ministers at Leipzig in 1994.
- During the French Presidency, trend scenarios for the territory of the Union in the year 2015 were conceived. These trend scenarios were the result of the contributions from all 15 Member States and were presented at the ministerial meeting at Strasbourg in 1995. They concentrate on three "Spheres of Activity" identified at the earlier Leipzig meeting: urban systems in Europe, infrastructure and transport and natural and cultural heritage.
- Under the Spanish Presidency, the Member States and the Commission were asked to cooperate in elaborating the trend scenarios. This resulted in a list of problems and solutions for each Member State. At the Madrid meeting in 1995 four criteria for identifying European spatial problems were defined:
  - economic integration
  - spatial integration
  - geographical location: centre/periphery
  - restrictions/possibilities for sustainable development
- Preparatory work on the ESDP continued and was elaborated upon during the Italian Presidency, when, for the meeting at Venice in 1996, a survey of integrated management and development of the cultural heritage was conducted. Another survey covering cities and metropolitan regions and focusing on globalisation, sustainability and a new view of land use and future spatial developments was also conducted in preparation for this meeting. Furthermore, attention was paid to maps. Amongst other things, an effort was made to capture the Madrid criteria on maps and, finally, an inventarisation of all possible policy options named by the Member States was conducted.

After this period of reflection, it was decided to collect all these ideas in an initial consensus document. This "first official draft" of the ESDP was presented at Noordwijk in June 1997.

5 Contents

The structure of the ESDP is illustrated in Figure 1. Of the four parts that make up the total document, most attention in this chapter will be focused on Parts II and III. Part I forms an introduction to the context within which the document must be viewed and provides general information. Part II forms the basis on which spatial planning decisions must be made and Part III represents the heart of the document. Here, the actual policy aims are defined. An outline of Part IV, the implementation strategy, is given in the first official draft of the ESDP. Work to complete this part will continue in the future. Its contents will be dealt with in Section seven of this chapter.

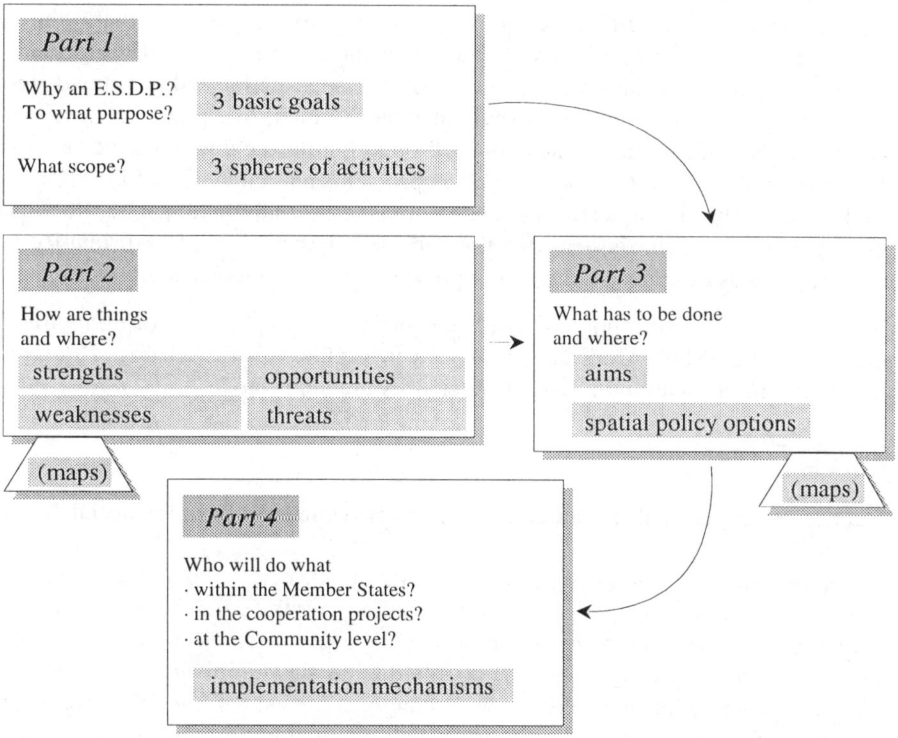

Figure 1. The structure of the ESDP

In Part I of the ESDP the broader social and economic context is described. The necessity of composing and formulating European spatial developments, and the surplus value the ESDP generates when translating these developments into policy aims, are important elements. Also dealt with are the status of the document, its basic goals, and the guiding principles on which the ESDP is built.

5.1 Analysing Territorial Issues

As a means of providing a description of the European territory, the forces driving the long-term territorial development of Europe are analysed in Part II of the ESDP. The main strengths and weaknesses combined with the most significant opportunities and threats are identified. As the locational pattern of activities and of land use within the individual Member States is increasingly being influenced by external developments, it is necessary to provide a Europe-wide overview of such developments and issues.

"Three elements are paramount in this overview:
Firstly, European territorial issues are above all, but not solely, those that occur at continental and transnational level. They also include issues manifesting themselves at regional or local level. At these levels, issues are located at many different locations within the Union. Common interests then give rise to a European dimension. Secondly, such issues point up opportunities arising from the specific strengths of the overall territory as well as problems and threats resulting from its weaknesses. Thirdly, a territorial approach is essentially a dynamic, forward looking approach. The issues identified in part II therefore relate primarily to emerging trends as opposed to existing present-day situations" (ESDP, 1997).

Fundamental starting points have been formulated for the territorial approach targeted by the ESDP. Alongside the basic geographical characteristics of Europe, the driving forces within the European territory are delineated.

Driving Forces

The ESDP delineates three major driving forces behind trends in territorial developments in Europe.

The first of these forces is composed of the major demographic and urbanisation trends. The relatively low rate of population growth (< 0.5 %) will pose new challenges such as how to maintain sufficient levels of services at the regional level. Furthermore, the ageing of the population, i.e. the diminishing proportion of younger people, is leading to rising and changing demand for housing. Moreover, the main effect of growing migration on both population growth and distribution is to reinforce population growth in larger urban areas, where migrants tend to settle the most.

The second group of trends marked as a driving force for the European territory are the typical characteristics and trends of the European economy. These trends will shape new economic geographical patterns. Examples of trends are the distribution of larger, medium and small enterprises, the internationalisation of trade within Europe, the introduction of the EMU and the further process of economic liberalisation and new communication technologies. These trends, or market forces, have the potential to steer developments towards a greater imbalance of the European territory. It is therefore important to understand them and identify the opportunities they can open up.

Environmental trends are the third driving force identified in the ESDP. There are six major environmental issues that are named in this respect:

- waste reduction and management
- climate change
- loss of biodiversity
- acidification and pesticides
- freshwater management
- forest degradation.

The ESDP states that the fundamental question in tackling these issues at the roots involves how to devise policies of a more structural nature to address environmental trends in the long term. This longer term approach is important as territorial strategies can play a significant role in implementing environmental policies.

Territorial Issues

When analysing the European territory, the range of issues that are relevant is practically infinite. By observing where the opportunities as well as the threats for the European territory have their roots, policy priorities can be formulated accordingly. In the ESDP therefore, only issues of European significance have been set forth. Issues are classified in line with the subject areas covered by the guiding principles, to wit urban structures, the role of rural areas, transport, communication and infrastructure and the pressure on the natural and cultural heritage.

When regarding the urban structure of the European Union, it becomes obvious that it is one of the most urbanised continents in the world. This structure is a mirror of national urban developments and accordingly consists of a relatively large number of major, international cities spread throughout the continent. These cites are in turn linked to small and medium-sized towns that are situated relatively close to one another. However, as a consequence of the economic, social and technological changes in society, a change in the hierarchical urban system is discernible. Nowadays the urban structure is alive not only to changes within the national context, but also to the competitive demands and challenges of the whole territory including the opening up of central and eastern Europe. An example is the way in which Berlin is being transformed from a divided, isolated city into a major capital gateway between western and central and eastern Europe.

The most important territorial issues that arise from these developments in the urban structure are the emergence of networks and urban clusters, changing urban economic opportunities, continuing urban sprawl, increasing social segregation in cities and inadequate improvement in the quality of the urban environment. The emergence of urban clusters, engendered, for instance, by increasing cooperation between cities, can stimulate social and economic cohesion in border areas.

In the past, the structure of rural areas was to a large extent determined by the Common Agricultural Policy. However, there is now a transformation underway from a monosectoral to a more integrated type of policy in which a territorial approach has an important part to play. This change can be regarded as a fundamental change in the position and function of rural areas in Europe, according to the ESDP. Two major territorial issues can be identified as a consequence of these changes. Firstly, there is an increasing interdependence with urban areas. Secondly, a shift can be detected away from agriculture as the economic base of rural areas.

In the European Union, changes in the transport, communications and knowledge infrastructures are taking place on a large scale. As infrastructure to a large extent determines the organisation of Europe, these changes have major impacts on its territory. Generally speaking, a tendency towards concentration and polarisation can be identified. Some of the consequences of these developments are unresolved border inconsistencies, increasing transport flows, increasing congestion and bottlenecks, a tendency towards concentration and emerging development corridors and disparities in the diffusion of information and knowledge. Increasing transport flows, for instance, are a result of the enlargement of 1994, the completion of the single market, and the opening up of central and eastern Europe. As the existing infrastructure is still very much nationally orientated, attention is being paid to the evolution of transport and infrastructure policy at the Community level.

The continuing pressure on Europe's natural and cultural heritage needs attention at a European level. Even though this pressure poses threats to the heritage, increased attention can create opportunities as well. Issues that are named in this respect are first of all the loss of biodiversity and unspoilt areas. The natural diversity that characterises Europe is a significant asset that is gradually diminishing. Secondly the continuing wastage and pollution of water poses a threat to the natural heritage throughout Europe. Water management can prevent problems regarding water distribution as they exist at the moment. As agriculture is one of the largest water polluters, agricultural policy can be expected to take more account of the water pollution it initiates. In this way positive results can be achieved at the European level. By focusing on spatial planning capable of preventing clashes over land use, pressure on cultural landscapes, of which Europe has a large diversity, may be relieved. Awareness of the large diversity of cultural landscapes that Europe has is crucial. As well as the rural heritage, pressure on Europe's urban (cultural) heritage is an issue here as well. This heritage is of enormous historical, aesthetic and economic value to local, regional and national communities. Next to the efforts already being made, an integrated multisectoral strategy is needed to safeguard it.

Existing Community Policies

As there is no specific spatial policy at European level, existing sectoral policy with a spatial dimension has a relevance for spatial development at a European level. Nevertheless, many existing Community policies do have a significant impact on territorial development in Europe. Besides this, Article 130A compels all Member States and the Commission to contribute to the objective of economic and social cohesion. One of the acts most conducive to an increase in the authoritativeness of EU policy was the signing of the Maastricht Treaty. The treaty brought a notable extension of powers for sectoral policies with a significant territorial dimension. The impact of these policies can be measured by applying a number of criteria. The following criteria are named in the ESDP:

- effect on land-use practices
- effect on the economic structure of regions
- effect on the spatial structure (urban system, infrastructure)
- effect on income and productive capacity
- regional support measures
- spatially differentiated public support measures.

The policies that potentially have the most territorial impact according to these criteria are the Structural Funds policies, the Common Agricultural Policy, transport policy and Trans-European Networks, the RDT policies and Community legislation in the environmental field. Of these, the CAP and Structural Funds are the most important in financial terms.

The ESDP is an attempt to coordinate the impact of existing EU sectoral policies affecting the European territory as a whole. "Member States and the Commission consider spatial planning as an important contribution towards raising the efficiency of the various sector policies of the EU" (ESDP 1997).

5.2 Policy Aims and Options for the European Territory

For each guiding principle of the ESDP a selected number of policy aims have been defined in Part III. These aims are backed by a number of policy options. The options can be identified as stepping stones to help achieve these aims. They are to be seen purely as suggestions for future policy. Each option can be associated with a list of geographical items.

A set of general criteria has been used in the selection of policy aims and options:

- the territorial dimension: policy aims and options should complement sector policies and, in line with the integrated, multi-sectoral approach of the ESDP, function as the modules of this approach.
- the European dimension: the policy aims and options should address issues that cannot be solved at a national or regional level.
- interest of the Member States: the Member States have expressed their priori-

ties in earlier ministerial meetings. The aims and options need to reflect these priorities.
- feasibility of implementation and potential effectiveness: the aims and options are also selected on the basis of these criteria.

More Balanced Polycentric System of Cities and a New Urban-Rural Relationship

Changes in the urban system and patterns of settlement are wide-ranging. Europe's cities and the way in which their inter-relationships are changing is very significant for European cohesion and competition. A polycentric organisation offers new prospects for the European territory. It must be noted that the word polycentric used in this context refers to a balanced pattern of cities, and not to a pattern of cities of the same size evenly spread over Europe. When tackling issues relating to towns and cities, it is no longer possible to take a single urbanised centre and its hinterland into consideration. The broader context, in which urbanised centres are seen as nodes within urban networks, needs to be taken into account as well. It is necessary to consider the urban structure of a territory, be it at a regional, transnational or European level. A polycentric structure enables cities to interact with each other in complementary and competitive senses and provides the periphery with new perspectives. In addition, the environmental quality of cities and social well-being in cities are parameters of their economic vitality. In the long run only sustainable cities will be competitive. Next to the changes occurring in the relationships between cities, the relationship between the urban and rural structure of Europe has also entered a new phase in which the diversification of rural economies and the development of higher functions in cities that are fragile nodes in the European urban system are central elements.

With these changes in mind, complementarity and cooperation between towns and cities is the first of the aims addressed in the ESDP. Connections between cities are important as a means of improving the balance of urban systems in Europe. The contexts in which this aim can be achieved may be in, for example, clusters of cities or in urban networks at various levels of magnitude - regional, transnational or even European. Cities can react to the aim of complementarity and cooperation in various ways. Cooperation, for example, can be based on certain themes that cities have in common such as social exclusion, or cooperation can go as far as integrating common goals. In order to address the imbalances of the urban system, physical and functional interlinkages are necessary. Examples of imbalances are polarisation around a central metropolis or the tendency towards ribbon development.

The second aim here is "dynamic, attractive and competitive towns and cities". Towns and cities are the motors of a region's economic growth and the momentum of their economies must be ensured. As a consequence of the increasingly global and liberal economy many cities face competition previously unfamiliar to them. One of the possible changes this could create is the polarisation between successful and less successful cities. Cities addressed in this aim are those that need an extra internal boost before they can face up to this competition or else cities that need attention if they are to be able to accommodate the extra boost of the world economy, notable examples being "gateway cities" (e.g. major seaports, cultural centres, intercontinental airports) providing access to the Union's territory. Moreover, extra attention needs to be paid to small cities that can provide an important boost to the revitalisation of rural areas. None of these topics can be achieved by cooperation alone.

Another important factor with a European dimension is the sustainable development of cities. The economic, physical and social situations of towns and cities are important factors in the economic development of Europe. The existing integrated model of a sustainable city deals with these factors, and therefore attention should be paid to the sustainable city approach. The five features of the model that are relevant to spatial planning are: checking the expansion of cities, the mixing of functions and social groups within the urban fabric (particularly in metropolitan areas where areas of exclusion are arising), the prudent management of the urban ecosystem (particularly water, energy and waste), the development of modes of access which are both effective and environmentally friendly, and the conservation and development of the cultural heritage.

The partnership that exists between towns and the countryside plays an important role in the changes taking place. Technological and economic integration in Europe are engendering a greater differentiation of rural areas. In rural areas, therefore, local needs and circumstances should be taken into account where spatial development is concerned. The role that agriculture plays in different regions, with respect to the amount of land it occupies, for example, or the role it plays in the rural economy, can vary greatly. The changes taking place in the development of rural areas should be linked to a reassessment of the partnership between towns and the countryside. These changes are most prominent in rural areas that are subject to new forms of pressure from development and those that are affected by the decline of agriculture.

The diversification of rural areas and the role that agriculture plays in land use and the rural economy is an important aspect to take into account when regarding policy aims and options.

Two situations are likely: highly productive agricultural production can compete in open markets, while agriculture that is not competitive with its products needs to undergo conversion. Spatial development will have to take diversification into account in both cases. A particularly interesting source of diversification for rural areas is the potential for renewable sources of energy.

Parity of Access to Infrastructure and Knowledge

Infrastructure is, as mentioned earlier, one of the most important factors in the organisation of the European territory. A tendency exists towards concentration and polarisation of transport and information flows. This process could cause uneven development in the transport and communications sector and undermine social and economic cohesion. Measures must be taken to prevent attention being exclusively focused on the most profitable parts of the network. Steps also must be taken to combat the "tunnel effect" and the "pump effect", these effects being respectively the spatial concentration in corridors, and the possibility of moving to economically stronger areas when these become accessible to areas that have always been comparatively isolated. Investing in intelligence and challenging the intangible economy of the Information Society through the dissemination of knowledge and innovative capacity are also imperatives.

The first aim, "better accessibility" for the whole Community, not only requires the completion of the main Trans-European Networks (TENs), it also involves attention being paid to secondary links through, for instance, the promotion of efficient public transport and an all-embracing telecommunications sector. If the TENs are not completed and only the most profitable parts are put to use, the risk of polarisation is considerable. It is also worth noting that better accessibility not only involves the more peripheral and remote areas but also isolated areas within central regions. Moreover, issues of accessibility need to be viewed at an intercontinental level. Access to Europe from other parts of the world is not evenly distributed and similarly requires attention. Summing up, it is clear that development strategies, especially at transnational level, are indispensable.

A more efficient and sustainable use of infrastructure is the second aim mentioned here. The efficiency of traffic is being harmed by its own growth. Especially on routes located in the centre of Europe and along coastlines, congestion is becoming such a problem that it requires special measures such as road pricing. These measures can only be effective if they are coordinated. At the same time, the growth of traffic is causing environmental harm. A more sustainable use of infrastructure can be supported through regional planning and the integration of transport and spatial planning policy, a decrease in mobility, and a careful location of activities. Concrete measures include encouraging location policies that reduce dependence on private cars and improving links between regional and national transport services. The most important measure, however, and one that is essential to both the environment and the efficiency of traffic, is the promotion of multimodality. Coordinated strategies and cooperation between carriers are vital if complementarity between transport systems is to be enhanced. "For instance, two neighbouring airports could share a single rail link rather than building two separate ones" (ESDP, 1997).

When regarding the competitiveness of Europe, accessibility to information and knowledge is crucial. Regarding the future, it is probably even more crucial than accessibility to infrastructure. Therefore, the diffusion of information and knowledge is a policy aim. Regions that are able to meet the challenge of the Information Society will increase their capacity to be innovative and create new economic opportunities. For the diffusion of innovation and knowledge attention needs to be given to ensuring that all areas of the European territory are in a position to partake of the Information Society, be they remote or centrally located areas. High quality services at reasonable charges must be provided for all. The strategy to be followed implies raising awareness on this matter and stimulating the distribution of higher education and research facilities throughout Europe.

Prudent Management and Development of the Natural and Cultural Heritage

The natural and cultural heritage of Europe is very diverse. Both types of heritage form a large and important asset for Europe and her economy. In order to protect this heritage, it is not enough to simply pursue conservation strategies. A more appropriate strategy can be found in stressing the evolutionary nature of heritage, i.e. favouring a developmental approach. Today's achievements add to past heritage, thus strengthening the heritage that is passed on to the future. Attention to water as a natural resource and to the urban and rural landscapes deserve special attention. Opportunities and threats that are arising as a consequence of the enormous growth in the tourist industry are aspects to take into account when regarding Europe's heritage, both natural and cultural. Greater attention to careful management and development is required.

Regional characteristics require specific regional attention in the case of the natural heritage. All policies with a territorial impact should be considered in local conservation strategies. The ESDP defines three main types of area that can be distinguished: protected areas, sensitive but unprotected areas such as mountains, coastal areas or areas affected by urban pressure, and rural cultural landscapes. The number of protected areas has risen in the past few years, but is still split up into different separate parts. Attention must be paid to links between nature areas such as wetlands, valley floors and woodland. To ensure these links, the setting up of the Natura 2000 environmental network (habitat directive) must be concluded quickly. In this environmental network, areas that are of trans-European interest are to be linked to transnational and regional nature areas. Management strategies are needed to ensure protection and conservation of both protected and unprotected areas. Also, a sustainable approach to nature must be adopted to avoid the possibility of natural disasters that can occur through human intervention.

As mentioned earlier, water is a vital resource for practically all aspects of European life. In the future, however, supplying water could become a problem whilst demand for water will continue to rise. These difficulties might be of a quantitative as well as of a qualitative nature. The quality can be diminished due to, for instance, pollution, bad management, or overuse. Therefore, a water management strategy that addresses the problems of both quantity and quality is absolutely essential. Indeed, integrated strategies to manage water resources, flood or drought, are necessary. One aspect of water management that is easily overlooked is the fact that water can also be a danger and cause natural disasters.

The history of a region can be expressed through its cultural landscapes. One example are the "bocages" farmland landscapes along the Atlantic seaboard.

Cultural landscapes form regional identity and therefore they are a part of Europe's heritage and should be treasured. Protection of cultural landscapes presupposes sound management, respect for local conditions, good economic opportunities, and an integrated strategy. The creative guidance needed to protect this treasure should mainly be directed towards developing landscapes. The aim in this respect is to harmonise existing landscapes whilst adding contemporary developments in such a way that they will be an asset for generations to come.

The diversity of Europe's cultural heritage is not only expressed in its cultural landscapes, but also in its urban heritage. The great value of this heritage is a liability as well as an asset. Many historical sites are falling prey to degradation. Large investments are needed to safeguard their future. As in many other aims, a creative and integrated management strategy is what is needed to protect urban sites sufficiently. One aspect that should be mentioned is the fact that the urban heritage does not exclusively consist of old historical sites. Modern buildings or sites, such as the new modern centre of Lille or La Défense in Paris, can also be seen as additions to the urban heritage, in some instances at least. What is more, the entirety of social life and lifestyle in towns is an object that should be taken into consideration in conservation management strategies.

6 Territorial Analysis

In the wide-ranging discussion that the ESDP promotes, long-term guidelines for the European territory should be developed. In order to take a step in the direction of integrated spatial planning policies and to initiate a process of awareness with respect to spatially coordinated sectoral policies and measures, an analysis of the European territory is necessary. The ESDP has evolved a framework for the integrated application of the policy aims mentioned above together with the accompanying options. These aims and options can be identified on a territorial basis. This territorial framework embodies the added value of spatial planning at a European level in general, and the ESDP in particular. In this respect it is clear that on-going and more extensive research is necessary to ensure the quality and continuity of data on territorial issues.

Comparisons within this territorial framework of the strengths, weaknesses, opportunities and threats of sections of the European territory should be based on spatially relevant criteria. An agreement needs to be reached on these criteria and their indicators so that in the future typologies of areas can be developed and long-term guidelines be drawn up. In the ESDP, a first step has been taken in developing spatial criteria. Factors that play an essential role in this selection are the presence of one or more qualitative or quantitative indicators for each criterion as well as methods for combined assessment. Under Spanish Presidency, a set of four criteria were presented, and this has since been enlarged to seven. They reflect the multi-sectoral approach of the ESDP and could usefully form a platform for debate. In the period ahead, further research and analysis of these criteria will be necessary:

- geographical position
- economic strength
- social integration
- spatial integration
- land use pressure
- natural assets
- cultural assets.

Within this framework, three factors play an important role (figure 2). First of all geographical integration is a central theme. By this is meant the application of a combination of policy options to geographical entities. The second factor is horizontal coordination. This can be seen as a reference framework for the implementation of multi-sectoral policies for the area they cover. The policy options mentioned above can be applied to three different levels of magnitude: the European level, the transnational level and the regional/local level. Application of policy options complementary to each other at the different levels implies a need for vertical coordination, this being the third factor of the territorial framework. The territorial framework and the way in which these factors intermesh is illustrated in the following figure.

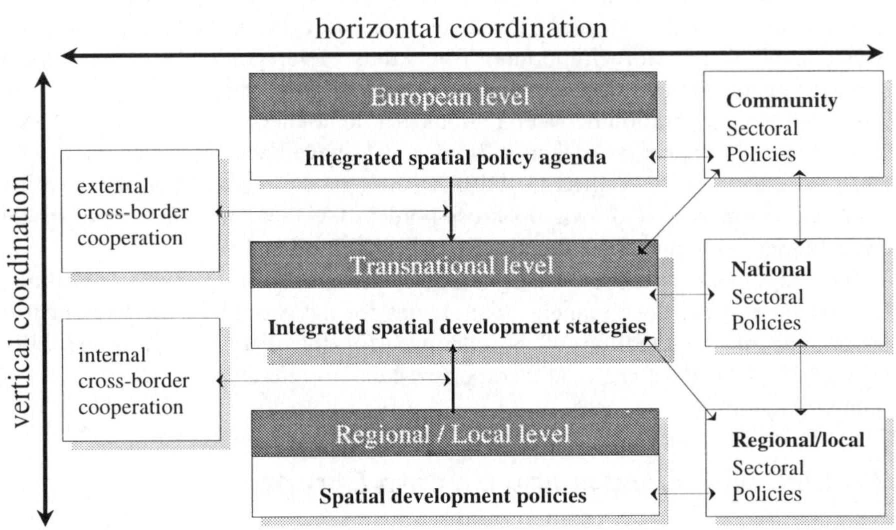

Figure 2. The system of integration

Policy coordination at the European level between spatially relevant sectoral policies such as transport, the environment and agriculture is already acknowledged. Interrelationships between policy options relevant to parts of the European territory could form a building block for the territorial framework and constitute an integrated spatial development policy agenda. A suggestion for such an agenda is made in the ESDP:

- a better territorial balance / economic and social cohesion
- balanced urban and regional competitiveness
- better accessibility
- reinforcing EU border regions and their cities
- the development of Euro-corridors
- sustain and develop European biodiversity.

Transnational areas usually embrace (parts of) three countries and are geographically contiguous. Complementary policies at European and regional/local levels are coordinated through the transnational level. This level therefore fulfils a "pivotal" role. Many European spatial issues are represented at this level. The INTERREG II C programme is an example of a transnational territorial programme. Seven transnational cooperation areas have been defined and Member States have expressed their willingness to cooperate on territorial issues such as flood prevention.

Most spatial planning policies are implemented at the regional and local level. European and transnational levels are often dependent on the regional and local levels as many spatial decisions are taken there. Integration and cooperation between levels is therefore important. Two kinds of integration in the European context can be defined, viz. development of cross-border spatial strategies and inter-regional cooperation between regions that do not necessarily share borders.

Regions that cooperate at a cross-border level have been supported by the EU since the creation of INTERREG in 1989. Amongst the regions that have profited from this support and drawn up cross-border development plans, thus giving cross-border cooperation a boost, are MHAL and the Saar-Lor-Lux.

Inter-regional cooperation mainly involves thematic policy options of common interest such as the development of strategies for the sustainable development of landscapes or the creation of programmes for the diffusion of accessibility through the linking of regional transport systems to interchanges at a national or international level.

7 Link with Central and Eastern Europe

The phrase "the balanced European territory" contains what must be the four most frequently used words in the ESDP. Needless to say, if taken literally, this area encompasses more than just the Member States of the EU. What many "West Europeans" fail to realise is that almost half of Europe lies outside the borders of the EU. Even though the ESDP is a perspective directed towards the territory of the Union, references to developments to the East of the Union's borders are also made.

The anticipated enlargement of the EU and the opening up of borders with central and eastern Europe have raised awareness of the fact that, once national borders are transcended, transcending intra-continental borders is the logical next step. Developments in eastern Europe will have consequences for the spatial development of the territory of the EU. The growing role of "gateway cities" such as Berlin or Vienna and the increase in trade and transport flows from East to West and vice versa are examples in this respect.

Globalisation and the growing interdependence of parts of the European territory that goes with it are trends that affect central and eastern Europe as well. Therefore, working together to cooperate on spatial planning and territorial issues is necessary for the whole of Europe's territory and not just within the EU.

There are a few conclusions that can be drawn with a view to future cooperation in the field of spatial policy in central and eastern European countries.

First of all, even countries that are used to working together, with similar planning cultures and a similar political framework, have needed a number of years to reach a consensus on a common spatial planning agenda. It is a gradual process and you cannot force your way through it. It takes time.

Another aspect that can be stressed is that the emphasis should be on improving existing policies rather than on trying to develop new ones. The aim is not to replace sectoral policy but to influence sector policies at the European level from a spatial perspective.

Finally, within the EU there is a distinction between Member States that have a strong tradition in spatial planning and those that do not. It is an important cultural aspect that needs to be taken into account when working together. It must be noted that this situation does not only apply for the EU. The cooperation process that goes with spatial planning aspirations on a wider and higher scale brings about an interesting exchange of policy-making cultures. This aspect gives the process a very important surplus value.

8 The Future

Conclusions of Noordwijk

The conclusions drawn after the presentation of the draft ESDP in Noordwijk can be seen as guidelines for the short-term future. The ministers of the European Union responsible for spatial planning have endorsed the document as the first draft of the ESDP. They have declared it to be an important step forward and have agreed to carry on the work in this direction. Besides this, clear mention is made of the non-binding status of the document. There is to be no connection between the ESDP and the distribution of Community funding such as the Structural Funds. It can be seen "merely" as a stimulus for policy makers.

It must be noted that at approximately the same time as the meeting in Noordwijk, an Inter-Governmental Conference was held under Dutch Presidency in Amsterdam. The basic thrust of the ESDP is towards further integration of European spatial planning following a set of guidelines. This is also the thrust of the Treaty of Amsterdam, which seeks to stimulate European integration in a broader context. Both conferences reflect the fact that European integration is a process in which progress is made taking one small step at a time. However, the link between spatial planning and the Treaty of Amsterdam has not directly been made during the IGC.

In a nutshell, the three main issues resolved in Noordwijk are:

- a broad political debate about the document will take place. The debate is to start as soon as possible and to end in the first half of 1998 under British Presidency;
- this debate will culminate in the first ESDP, to be presented in June 1998 at the end of the British Presidency;
- a working group will be initiated to set up an Observatory consisting of a network of European spatial planning institutions and to help refine and update ESDP analysis.

Besides these main issues, the conclusions of the Presidency in Noordwijk stress the importance of involving neighbouring countries outside the Union in the ESDP process. Collaboration with the Council of Europe and utilisation of existing Community programmes are both means of stimulating this involvement according to the conclusions.

Implementation

After the presentation of the first draft of the ESDP, strategies are to be drawn up to facilitate implementation of policy aims and options. These strategies are described in Part IV of the ESDP. Experimental and innovative initiatives will be undertaken to coordinate spatially relevant sectoral and territorial policies.

In the first phase of implementation, existing instruments of the European Union can be used. Among the existing instruments are the INTERREG II C programme for transnational projects and INTERREG II A, which addresses itself to cross-border cooperation policies. Next to these programmes, which were among the first to court territorial policies directly, Article 10 of the ERDF promotes pilot projects for the implementation of spatial strategies. Whilst the European PHARE, TACIS and MEDA programmes are not explicitly set up for the purpose of land use planning, they do supplement the existing territorial programmes.

Of course, implementation involves more than simply promoting programmes such as those mentioned above. Having described various policy aims and options in the ESDP, it is important to examine how these can be best implemented. The existing community policies are, however, a first step in the right direction. Through the existing programmes, experience will be gained and a basis for implementation of the aims can be achieved.

During the drafting of the first ESDP, gaps in information and a lack of comparable data were identified. When drawing up a document such as the ESDP it is imperative that sufficient data and information are available. The decision was accordingly taken at Noordwijk to set up an observatory network acting under the auspices of the CSD. This is not a new idea, since the wish for such an observatory was expressed as long ago as 1994 during the Leipzig meeting. However, at that time it was not quite clear within which framework the observatory should function and what exactly its tasks would be. Now that the policy aims have been defined and a first draft has been presented, a clear picture can be formed of what the tasks of the observatory will be. Two fundamental aspects of the observatory's tasks are stated in the ESDP. First of all its work should not conflict or overlap with other specialised institutions operating at a community level such as Eurostat. Secondly, it should not become dominated by bureaucracy.

Many players are involved in the field of spatial planning. However, when building a spatial planning perspective for the European Union, not all players can be included. Spatial planners have thus taken the step of authoring an ESDP concept and arriving at a consensus as to its contents. Now it is the turn of the other players to give their opinion and to likewise arrive at a consensus. The decision to hold a broad public and political debate in the second half of 1997 and the first half of 1998 gives all those involved the opportunity to air their opinions about the document.

In order to reach all parties involved, the debate should be held in a number of different settings; within each Member State, in the framework of the cooperation programmes, at the European level and, last but not least, with Member States or relevant international organisations outside the European Union. This should lead to the first official ESDP being drawn up and endorsed by the Ministers responsible for spatial planning. Issues that could be relevant to the debate can be divided into 4 categories, all relating to the territorial framework. First of all issues relating to horizontal coordination at the various levels of magnitude are mentioned, paying attention to how spatial policy makers can incorporate spatially relevant sector policies. Secondly, vertical coordination focuses amongst other things on how complementarity between top-down and bottom-up processes can be achieved. Geographical integration, being the third aspect of the territorial framework, can address the question of how to transcend the various administrative boundaries that exist within the European territory at all levels. As mentioned above, community policies, the fourth category, can help set off the implementation process. This is the reason why these policy instruments should also form part of the public debate, with the question needing to be asked as to how such instruments can be adapted to new tasks.

Only the future can teach us what the ESDP will give rise to. Whether or not spatial policies will play a meaningful role in the European spectrum remains to be seen. What the ESDP has already achieved is, as the Dutch minister for spatial planning Ms. Margaretha de Boer mentioned in her opening speech at the Noordwijk conference, that it has made both the spatial dimension of European policy and, conversely, the European dimension of spatial policy visible. The new context of spatial planning set out in the ESDP, a context transcending national and other administrative boundaries, indicates the progressive approach taken by spatial planning in Europe. New social, economic and technological trends in European society cannot be ignored by spatial planners, and the ESDP is a sign that spatial planning has made an effort to adapt to these new trends.

References

European Spatial Development Perspective; First official draft presented at the meeting of ministers responsible for spatial planning of the Member States of the European Union, Noordwijk, June 1997

Presidency Conclusions of the Informal Meeting of Ministers responsible for Spatial Planning Noordwijk, final version 2, 10 June 1997

Part II

Sustainable Development in Central and Eastern European Regions

Transfrontier Cooperation in Spatial Planning at the External Border of the European Union

Viktor Freiherr von Malchus
Akademie für Raumforschung und Landesplanung
Academy of Spatial Research and Regional Planning
Hanover, Germany

1 Odessa Oblast – Part of a Superregion Created by European Cooperation

Until recently, the Danube and Black Sea regions benefited from only limited international cooperation in spatial planning. Since the fall of the iron curtain in 1989, however, central and eastern Europe have moved into the public eye. Numerous central and eastern European countries wish to join the European Union, and most are now members of the Council of Europe.

The Council of Europe has been championing the idea of European spatial planning and cross-border cooperation for several decades. The Council of Europe is still striving to ensure greater attention is paid to spatial planning in international politics, particularly in central and eastern Europe, and, in conjunction with the German Federal Ministry for Regional Planning, recently organised a conference on "Sustainable Spatial Development for the Black Sea Region" in the Ukrainian city of Odessa.

The conference came at an opportune moment, coinciding as it did with a new Community initiative, INTERREG II C, launched by the European Commission in May 1996 to promote measures for sustainable spatial development in European superregions on the basis of cross-border cooperation. The states and regions participating in the INTERREG II C Community initiative (96/C 200/07) are currently drawing up an "Operational Programme for the Central, Adriatic, Danubian and South-Eastern European Space", in effect a European superregion which will also include Odessa-Oblast in Ukraine.

It is intended to provide a summary description of the emergence of cross-border cooperation in spatial planning since the Second World War, and indicate likely future developments.

2 Cross-Border Spatial Planning as a Component of European Regional Development

2.1 Important Measures Taken by the Council of Europe

Ever since its creation in 1949 the Council of Europe has been a champion of European integration and the solving of spatial planning problems in border regions. Such problems are addressed in the Council's institutions, i.e. the Parliamentary Assembly, the Committee of Ministers, and the General Secretariat, these being responsible for concluding agreements, issuing recommendations and developing strategies for joint procedures at practically all levels, including questions pertaining to spatial planning in cross-border cooperation. Cooperation with Council of Europe member states and territorial authorities is conducted at a variety of levels - parliamentary, regional/local, and, in the case of intergovernmental cooperation, ministerial. At all three levels, spatial planning problems and border region issues have been addressed as matters of particular urgency for the past twenty years, and measures have been proposed to promote and improve cross-border spatial planning.

Since the 1960s, the following major activities have been undertaken:

- together with the Standing Conference of Local and Regional Authorities of Europe, the Parliamentary Assembly of the Council of Europe has convened six "European Conferences on Border Regions", to wit in 1972 (Strasbourg), 1975 (Innsbruck), 1984 (Borken), 1987 (Saragossa), 1991 (Ovangemi) and 1994 (Ljubljana). The working party on European border regions was actively involved in preparing and running these conferences.
- over the past 30 years, the Parliamentary Assembly of the Council of Europe has adopted numerous recommendations pertaining to the promotion of cross-border cooperation and spatial planning. Since 1980, the main basis for these recommendations has been the European basic convention on the improvement of cross-border cooperation between territorial authorities. This convention has now been ratified by most members of the Council of Europe. On the basis of this convention, implementing treaties have been concluded between the BENELUX countries (1992), Germany and the Netherlands (1992), Italy and Austria (1993) and France and Germany, to name but a few; these treaties give local councils the statutory power to conduct direct cross-border cooperation.
- many proposals made by the European Conference on Border Regions and by the working party on European border regions have been incorporated in recommendations adopted by the Council of Europe's European Regional Development Ministers' Conference and found their most important expression in the European Charter on Regional Planning (1983), though in the definitive version of the Charter border regions and other territorial categories are relegated to the annexe.

The European Charter on Regional Planning stipulates explicitly, however, that the need of border areas for a policy of coordination and cooperation between nation states and regions exceeds that of all other regions. The Charter states that the goal of such a policy must be to open up the border, establish cross-border consultation and cooperation, and bring about the joint use of infrastructure.

As the meeting in Odessa showed, the Council of Europe will continue addressing cross-border issues in regional planning. For example, a seventh European Conference of Border Regions is being considered for 1997 or 1998, and the eleventh and twelfth Conferences of European Regional Development Ministers (CEMAT) in Cyprus (1997) and Hanover (2000) will also deal with the question.

2.2 European Union Activities

Since the mid-1970s, several reports dealing with cross-border issues have been adopted by the European Parliament's Committee on Regional Policy, Regional Planning and Relations with Regional and Local Authorities. All these reports, together with the proposal for a European Regional Planning Charter of 21/11/1983 address the question of cross-border regional planning in a variety of different ways. The European Parliament continues to believe that border regions are especially important for all the different dimensions of the building of the European Union, and in particular the Union's closer relations with the new democracies in central and eastern Europe and the promoting of cooperation between peripheral regions and third countries.

Over the past ten years the European Commission has been increasingly focusing on the question of cross-border regional planning and policy in EC policy. Noteworthy initiatives include:

- support for the setting up of "cross-border action programmes and cross-border development and action programmes",
- declarations on the development of border regions in the EC report Europe 2000,
- approval of INTERREG I and II (1990–1999), European Community Initiatives for border regions, and the drawing up of guidelines for Operational Programmes highlighting the importance the European Commission attaches to solving problems in border regions.

The objective of the INTERREG Community Initiative is to assist border regions in coping with specific regional planning and economic problems arising from their singular geopolitical status, to promote cross-border cooperation, and to prepare regions on the external borders of the Union for their future role in a Community characterised by greater integration. For the period 1994–1999 a total of 59 operational programmes have been submitted and approved under INTERREG II A, including 18 programmes covering areas on the EU external border with central and eastern Europe.

The Commission feels particularly obliged to promote cross-border measures involving the ex-communist countries. Since 1994, the Community budget has allocated an annual 150 million ECU of Phare funding for the promotion of cross-border cooperation (CBC) between central and eastern European countries and neighbouring regions in the European Union. For the future, it is important to establish a genuine link between PHARE CBC and INTERREG II A and to favour joint measures for PHARE countries and member states. Thought must also be given as to how PHARE, TACIS and structural funds can be used in tandem for projects involving regional and spatial planning which promote European unification.

On 8 May 1996, the European Commission adopted the INTERREG II C Community Initiative for transnational cooperation in the field of regional planning. The initiative is designed to promote balanced regional development in the European Union through structural measures contributing to the consolidation of economic and social cohesion and to sustainable and environmentally-sound development. It will be interesting to see how these new community initiatives unfold in the coming months. Member states with plans for joint transnational cooperation in regional planning for their superregions had finalised their Operational Programmes by the beginning of 1997.

3 Regional Development Planning through the Agencies of Interregional and Cross-Border Cooperation

3.1 Interregional Cooperation

The new language of European politics, now to an extent tailored to the requirements of European aid programmes, draws a distinction between "interregional cooperation" and "cross-border cooperation" in the narrower sense. Interregional cooperation can be defined as trans-European and transnational cooperation between state, regional and local players (regional and local authorities) and semi-public and private players (universities, regional development agencies, research institutes, social partners etc.). One must differentiate here between two forms of cooperation: multi-issue cooperation (when related issues are addressed) and issue-led cooperation (based on an external or a specific-sector approach).

The starting point for multi-issue interregional cooperation, examples of which can be found in many parts of Europe, tends to be the environment, research and infrastructure collectively. Only recently have specific issues involving regional planning and development been addressed.

The European Union recently launched two major projects for the promotion of regional development in Europe. Existing regional development concepts adopted by the European Parliament as well as the European Commission's "Europe 2000" strategy paper proved unsuitable for a full assessment and consideration of post-1989 developments in central and eastern Europe. As a result, during their informal meetings in Lisbon (1992) and Liège (1993), European Union regional planning ministers agreed to:

- develop concepts for the whole of the Community in the form of an extended "Europe 2000+" strategy paper,
- upgrade the role of Trans-European Networks and city networks in regional planning policy, to reflect their importance as planning instruments,
- advocate the immediate creation of an informal "Regional Development Committee" by the Commission,
- accelerate the drawing up of a European regional development concept based on member state cooperation.

The member-state regional development ministers argue that the master plans for future regional and communal planning in Europe should be drawn up jointly by member states and the Commission and be underpinned by the elaboration of a joint strategy. Since September 1994, the Leipzig document on the principles of a European development policy, drawn up by EU member states in conjunction with the European Commission, has furnished the political basis for further cooperation.

Since then, the Committee for Spatial Development (CSD), created in 1994, has continued working on a European spatial development concept in line with the principles of the Leipzig document. A further module in this concept has now been forthcoming in the form of a collection of development scenarios for European spatial planning in which an initial urban, transport or environmental planning scenario is posited, together with the salient development trends. These scenarios are not so much intended as portrayals of some ideal future form of spatial development in Europe but rather as an account of the challenges facing a future European spatial planning and development policy. Existing analyses and conclusions must be consolidated, preparing the way for a joint vision of aspirations for spatial development in Europe, on the basis of which effective spatial policies can be designed and implemented by the Union and member states. Regional, transport and environmental policy must be a priority. An initial draft version of a European spatial development concept should be ready by early 1997, during the Dutch presidency of the EU (compare van Ginneken).

A further major EU project is the already mentioned INTERREG II C Community Initiative, designed to extend spatial development in Europe with EU support. On the basis of Commission reports (Europe 2000 and Europe 2000+), the conclusions of eight informal meetings of European Community spatial planning ministers and studies of European sub-regions published in the EC document "Europe 2000+", the Commission has resolved to support spatial planning in geographical super-regions (for example the North Sea and Baltic regions) provided at least three member states are involved in such cross-border programmes. For example, various operational programmes for transnational cooperation in spatial planning are currently being prepared, within the framework of the INTERREG II C Community initiative, in super-regions in the east of the Union in conjunction with central European states. This Community undertaking has brought national and regional governments together, enabling aspects of spatial development on a larger scale to be discussed, analysed and wherever possible backed, which has to be a positive development. This has provided EU spatial development policy with new impetus. These operational programmes will provide for the investigation of measures for the preparation and implementation of transnational strategies, the introduction of measures to improve the impact of Community policies on spatial development together with measures to improve the spatial management of coastal areas at the periphery of the Union, including measures to enhance coastal regions. One particularly positive element of this Community initiative is the fact that special priority is to be given to measures for improving transnational cooperation in the field of spatial planning with third countries. These include the measures referred to above pertaining to the Danube and Black Sea region as well as those concerning the development of the Baltic region.

One example of interregional cooperation is the initiative of spatial planning ministers from the Baltic region. On 21 August 1992, spatial planning ministers from 11 countries bordering the Baltic met representatives from international organisations (EC, Nordic Council of Ministers, etc.) in Karlskrona, Sweden. Against the backdrop of a strong sense of unity in the Baltic region and a powerful common identity forged by historical events over recent centuries, the decision was taken to devise a spatial planning master plan for the Baltic Sea Region (BSR) in the form of a spatial development plan of action dubbed "Vision and Strategies around the Baltic Sea 2010". Its objectives include:

- promotion of the Baltic Sea Region internationally;
- support for the development of cooperation networks in the region;
- creation of a practical forum for cooperation and transfer of know-how;
- piloting of internationally important infrastructure projects;
- promotion of all areas of spatial development planning.

From the outset, it was decided that the development concept would not be a binding "megaplan" and that strategies and projects would only address issues affecting the whole Baltic Sea region, or at least such with a transnational dimension.

Upon completion of the project architecture in 1995, a series of major activities were planned or implemented, the first examples of joint action within the framework of this Baltic cooperation initiative. These include:

- conferences for BSR spatial planning ministers;
- elaboration of schemes, projects and pilot projects;
- PR for the BSR;
- construction of a research network and elaboration of research programmes;
- extension of the VASAB 2010 secretariat in Gdansk, Poland.

It will be interesting to see how these projects launched by the Baltic Sea Region spatial development ministers, in particular the "Visions and Strategies around the Baltic Sea 2010" spatial development plan of action, can be combined with the INTERREG II C scheme for the Baltic region. Both programmes emphasise the importance of cross-border spatial development planning.

3.2 Cross-Border Spatial Development Planning

Cross-border spatial development planning involves spatial planning cooperation between neighbouring states and territorial authorities responsible for border regions, as well as direct cooperation between municipal and local authorities on either side of the border; such cooperation can be at state, commune/regional, and local level. The function of cross-border spatial development planning is to address itself to marshalling the impact on spatial development of, respectively, society, the economy, and the natural, built and social environments in border regions.

An important factor for cross-border cooperation in the field of spatial planning were the bilateral and trilateral government and spatial development agreements of the 1960s, 1970s and 1990s, such as those in the Benelux countries and those involving Germany/Netherlands, Germany/France, Germany/Switzerland, Austria/Germany and Germany/Poland. Thanks to these agreements, governmental, spatial planning and regional commissions were created, which have conducted fruitful cross-border cooperation in the field of spatial development planning over the past thirty years. Their primary task is to provide a long-term basis for established forms of cross-border cooperation between authorities at the different planning levels, in compliance with the principle of subsidiarity. Cooperation between Germany and Poland provides an excellent illustration of this process in central Europe; in recent years, governmental and spatial planning commissions have created a "spatial planning model for German-Polish cross-border cooperation" and a handbook of planning terminology.

The spatial development policy model for the German-Polish border region provides a blueprint for the desirable future development of the cities, communes and regions on either side of the border. State and local authorities on both sides of the border are responsible for the implementation of this model. As a result it furnishes a good example of cross-border spatial development planning along the divide with central Europe. The German-Polish spatial planning commission was set up soon after German reunification on 2 July 1992. Its activities bring together the most important spatial planning bodies from the two countries. By 1995, the commission was able to publish the "Spatial Development Master Plan for the Region along the German-Polish Border". This document enunciates two principal objectives:

- improving living conditions, and in particular bringing living standards up to those of other regions;
- eliminating the barrier effect of the border by promoting cross-border communication and cooperation.

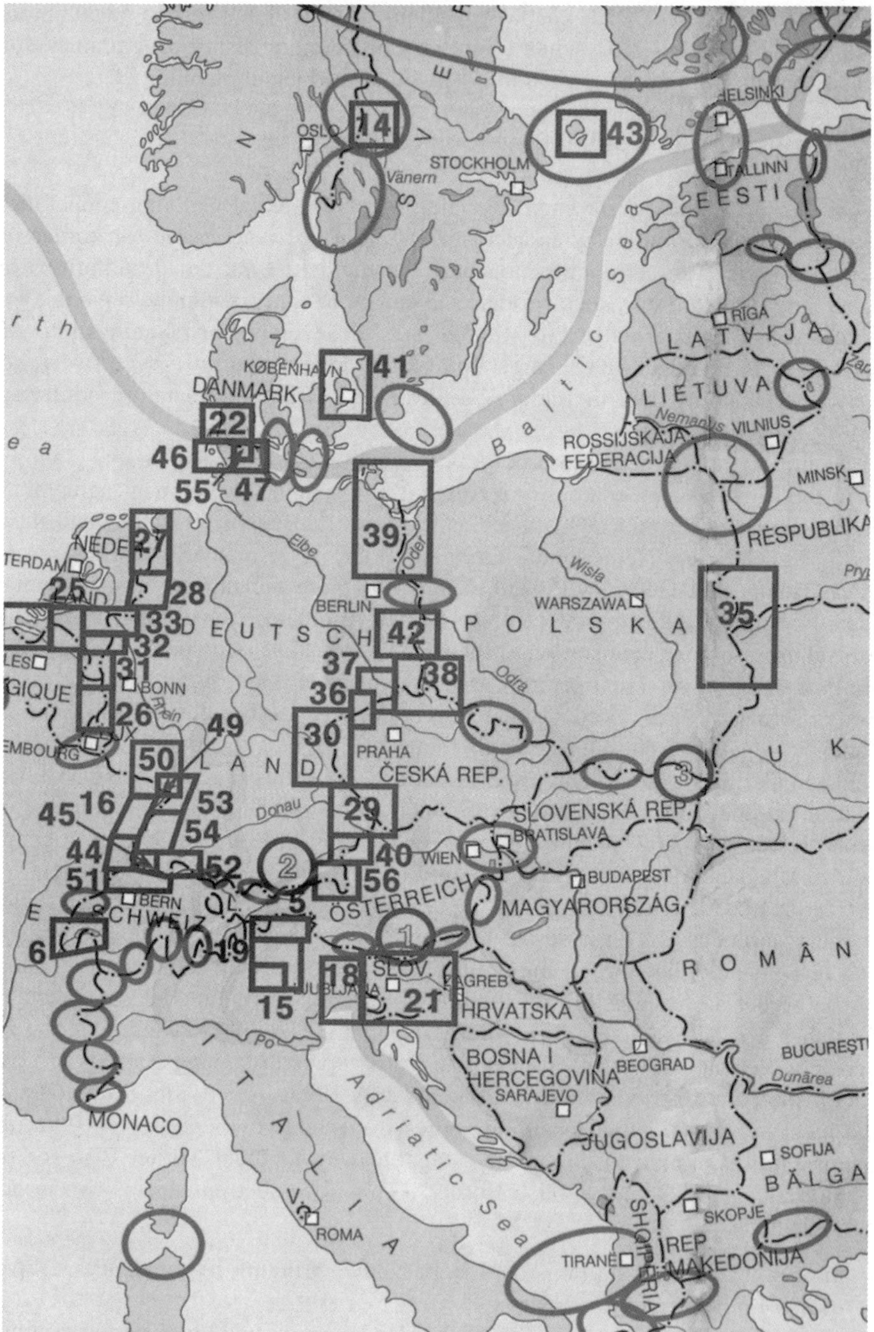

Figure 1. Border Regions and cross-border regions

Against the backcloth of these general statements of intent, the master plan devotes itself to three main issues:

- protection and enhancement of natural potential and the rehabilitation of damaged areas;
- development and nurturing of decentralised communal structures either side of the border;
- development and extension of cross-border technical infrastructure.

On 24 May 1995, the spatial planning commission adopted a resolution recommending that all planning agencies at national, regional and local level, on both sides of the border, include these master plans and objectives in their planning once compliance with the legislation of both states has been ensured, laying special emphasis on cross-border aspects.

Regional/local cooperation in spatial development has also been extended and deepened in recent years. Through INTERREG the EU has made a major contribution to the promotion of such cooperation. One important benefit of this assistance has been the setting-up of a total of eight "Euroregions" on the German-Polish and German-Czech (4) borders, one of which, the Euroregion Neisse, includes regions in both Poland and the Czech Republic. Euroregions have succeeded, for example, in increasingly integrating their own development ideas into overarching spatial planning models, thereby exerting a positive influence on state, regional and national spatial planning on both (or indeed all) sides of the border. It should be emphasised that Euroregions straddling the border with central Europe have worked in exemplary fashion with state planning authorities in recent years, thereby contributing to ironing out the bumps in spatial planning policies between neighbouring countries. Building above all on partnership and subsidiarity in planning, cross-border successes have been notched up thanks to:

- the elaboration of cross-border master plans with the participation of local and regional authorities;
- cross-border transport planning throughout Europe;
- the inclusion of spatial planning considerations and objectives in the cross-border development and action programme for regional policy and in the operational programmes of INTERREG I and INTERREG II.

Further intensification of cross-border cooperation in the field of spatial planning would also be desirable, including greater participation by local and regional authorities. Intensification of cross-border spatial planning must be facilitated in future by:

- accurate analysis with the help of scenarios of current and future spatial development problems;
- studies of the legal hurdles and restrictions hindering an intensification of cross-border planning cooperation;

- ongoing cross-border information procedures and spatial monitoring with the help of socio-economic studies of spatial development;
- the creation of cross-border spatial development master plans and schemes to facilitate synchronised macro-spatial planning for neighbouring regions;
- the identification of important urban regions (city networks) in border regions, primarily areas where major centres interconnect and cities are split in two by state borders;
- regular cross-border coordination of all spatial planning projects and measures;
- surveys to identify action zones for spatial planning policy in border regions;
- identification of nature reserves and interconnecting biotopes, cross-border natural parks and cross-border landscaping;
- the setting up of rehabilitation programmes to enhance the environment;
- special cross-border planning for border regions affected by the upgrading of major (trans-European) transport routes;
- coordination of local planning in all border communities.

To ensure optimum implementation, all spatial planning and regional policy measures should, in future, be embedded in "regional cross-border development plans of action" and regional economic policy "operational programmes". Good examples of such planning are to be found on the German-Dutch and the German-Polish borders.

It is important that cross-border cooperation be clearly structured and based on institutionalised forms, that community-led policies involve all stakeholders, and that local and regional politicians be encouraged to participate in cross-border activities.

The fundamental objective of cooperation both within and outside the European Union is the consolidation of peace. People cannot plan their future and play an active role in achieving cross-border objectives without genuinely stable and predictable conditions. Shared values – democracy, human rights, ideals of equality and the sustainable ecological development of border regions – are factors which unite the member states of the European Union. Interregional and cross-border cooperation along the internal and external borders of the European Union is called upon to deliver these EU ideals.

References

Akademie für Raumforschung und Landesplanung, ARL (1996) Europäische Raumentwicklungspolitik - Rechtliche Verankerung im Vertrag über die Europäische Union. Arbeitsmaterial Nr. 233, Hannover

Bundesministerium für Raumordnung, Bauwesen und Städtebau (1995) Grundlagen der Europäischen Raumentwicklungspolitik. Bonn Bad-Godesberg

Deutsch-Polnische Raumordnungskommission (1995) Raumordnerische Leitbilder für den Raum entlang der Deutsch-Polnischen Grenze – deutschsprachige Zusammenfassung. Bonn und Warschau

Europäisches Parlament, Generaldirektion Wissenschaft (1996) Die Grenzüberschreitende und Interregionale Zusammenarbeit in der Europäischen Union. Arbeitsdokument erstellt von der Arbeitsgemeinschaft Europäischer Grenzregionen (AGEG), Gronau, Luxemburg

Klein R (1995) Grenzüberschreitende regionalpolitische Zusammenarbeit im Rahmen von INTERREG und PHARE. In: EUREG, Nr 2, S 48–52

Kommission der Europäischen Gemeinschaften – Europäische Kommission, Generaldirektion XVI (1994) Europa 2000+ Europäische Zusammenarbeit bei der Raumentwicklung. Brüssel, Luxemburg

Krautzberger M, Selke W (1996) Perspektiven der bundesstaatlichen Raumplanungspolitik in der Europäischen Union, das Beispiel der Bundesrepublik Deutschland. In: Institut zur Erforschung von Methoden und Auswirkungen der Raumplanung der Ludwig-Boltzmann-Gesellschaft, Technische Universität Wien, Rudolf Wurzer (ed) Wien

Malchus V Frhr. von (1995) Europäische Raumordnung. In: Akademie für Raumforschung und Landesplanung (ed) Handbuch der Raumordnung, Hannover, S 261–268

Implementing Ecological Development Concepts in Border Regions
Project Experiences and Conclusions Drawn

Isolde Roch
Institute of Ecological and Regional Development
Institut für ökologische Raumentwicklung
Dresden, Germany

1 Introduction

Due attention was given to transnational cooperation at the EU's eastern border virtually from the moment the process of political reformation began in Europe (cf. Selke 1994, von Malchus 1994, Roch 1995, Moll 1994). Regional planners were notable in discerning the need for action as well as opportunities to proceed in a targeted and orderly manner with the objectives of ecologically sound development in the border regions in the process of being constituted. Building on the experiences of transnational cooperation in the Alps and in the Franco-German and Dutch-German border regions, where there have been exemplary successes at municipal level, preference was given in the case of cooperation at the external border of the EU to the elaboration of transnational development concepts of an interdisciplinary nature with regional planning at the helm. With the mutual consent of the neighbouring countries and before fully-fledged regional development plans had been put in place in the Czech Republic, Poland, or the resuscitated German states of Saxony or Thuringia, transnational regional planning strategies were fashioned, first between Bavaria and what was then still Czechoslovakia and a little later for the Saxon-Czech and the entire German-Polish border area, that constituted common goal programmes and orientation mechanisms for development at national and regional level as well as for the municipalities with their divergent outlooks, powers to act, and degrees of motivation as regards transnational cooperation. This across-the-board involvement of planning levels and players in informal cooperation right from the drafting stage and thereafter in programmatic implementation through project work proved to be extraordinarily propitious in defining where action was needed and what would be the way of environmentally sound development. By means of this open forum an understanding of the border region was gained.

The present contribution addresses itself to how development concepts are constructed with particular reference to the action needed and the objectives of development (Chapter 2) and to the current state of implementation of these concepts (Chapter 3) as well as supplying recommendations as to the further course of transnational cooperation from a specialist viewpoint (Chapter 4).

2 Construction of Transnational Development Concepts

It is intended to illustrate how transnational developments are constructed by citing the Trilateral Development Concept (Roch et al. 1995) in south-western Saxony, the territory of the Euregio Egrensis.

The transnational development concept for the region where Bavaria, Bohemia, and Saxony meet was drawn up by regional experts on behalf of the Bavarian State Ministry for Regional Development and Environmental Issues, the Saxon State Ministry for the Environment and Regional Development, and the Czech Ministry of Trade and Industry and adopted by the relevant ministers in 1994. For Saxony, it was the first case of policy being primarily oriented towards landscapes with organically evolved settlement structures as decreed for the elaboration of regional development plans in the Saxon Regional Development Act. The Bavarian approach to regional planning and the system of 'ecological stability' practised in the Czech Republic ably reinforce the ecological thrust of development planning for the region.

The plan of action breaks down as follows:

- inventorisation of regional structures and developments
- strengths, weaknesses, and where action is needed
- guiding principles and goals of development
- catalogue of measures and
- pilot projects with the task of initialising implementation of guiding principles and goals

The ecological assessment of the border area took the form of an exhaustive log and summary of nature, landscape, drinking and spa water conservation areas and soil quality together with a delineation of the state of protectable assets reflecting the degree of pollution of flowing waters, the atmosphere, and woodlands, damage to the latter being the upshot of noxious emissions stretching back many years that have had an extremely detrimental impact upon the quality of life in the regions affected and most tellingly upon the fortunes of the spa tourism industry.

Natural assets in particular need of conservation were dubbed "sensitive areas" to provide a spring-board for their joint preservation and development and for the general and economic well-being of a given area. This spatial representation of areas as a high-priority means of safeguarding and developing ecological functions is an important and exemplary result of the border area concept that points up the basics of the direction development is to take in both the border region and in subregions and municipalities. The specific requirements of environmental protection and biotope networking are to be catered for at regional and communal level within the scope of this safeguarding of the natural inventory, with emphasis also being placed on aesthetic items of note. At the same time, attention has to be given to the wide-ranging need for rehabilitation schemes to render ecosystems fully viable once again. The division into "inner" and "outer" sensitive areas indicates differences in the way areas are to be treated. Inner sensitive areas are entirely given over to conservation measures whilst their outer counterparts may be optionally used for other purposes as long as specific site-related provisions are met.

Drawing on the strengths and weaknesses of the border region, the following areas requiring action were specified (Roch et al. 1995):

- preservation and consolidation of the diversity and attractiveness of landscapes, partly with an eye to further developing the residential factor, tourism, and other location-enhancing features;
- maintenance and extension of the share of near-natural lands to be conserved;
- lowering emissions by closure or modernisation, by installing or modernising air purifiers, by foregoing fossil fuels in favour of more environmentally acceptable solutions and by evolving mutually acceptable transport strategies designed to cut levels of private motoring;
- introduction of rehabilitation schemes (e.g. to remedy woodland and erosion damage);
- cleaning up polluted bodies of water, above all by treating industrial and domestic effluent;
- protection of drinking-water sources;
- upholding the mosaic of small-scale land use;
- generally prioritising ecological issues as a means of improving the investment climate and living conditions.

The guideline for the ecological localisation of the border region was stipulated (Roch et al. 1995) as being:
Preservation and shaping of an ecological role-model and pilot region

Systematic use is to be made of the opportunity presented specifically by the common Bavarian/Saxon/Czech border area to

- conserve natural resources,
- sustain the fine fabric and variety of landscapes, and
- safeguard the common cultural heritage.

Wherever possible, existing ecological assets are to be preserved, forming as they do an increasingly vital aspect of qualitative spatial development. The yardstick for economic, residential, and infrastructural development schemes shall be their environmental acceptability.

This guideline underscores the earnest with which the region is pursuing ecologically-based modes of development both for industry and the domestic sphere. This political declaration of intent based on an analysis of strengths and weaknesses is also driven by an awareness of the growing significance of "soft" business locations for the stabilisation of economic development and the residential factor and also, no less importantly, for the fortunes of the spa and health resort sector. Discussions in the steering group and during demonstration events in the three regional centres at which the guiding principles and goals were adopted can be seen as constituting an initial debate on the fundamentals of a sustainability paradigm given this commitment of the border region to the principles of environmentally sound development.

The disciplinary guiding principles and goals established are backed up by measures intended to pump prime the development process that were proposed with the specific aim of playing to the border region's strengths and minimising its weaknesses. On the basis of action needed as well as the findings of a written questionnaire involving municipalities, villages, and local authorities in the Saxon border region, a catalogue of measures was compiled (Roch, Leimbrock et al. 1996) and presented to the border region as discussion material at the regional association meeting. An eminently fruitful discussion ensued that led to the proposals being added to and fine-tuned and revealed a distinct need for coordination in carrying out generically related tasks such as care of landscapes, structural redevelopment, or landscape rehabilitation. Momentous decisions were made concerning what were to constitute priority measures, whose implementation was expected to act as a trigger as well as having repercussions a significant way across the adjoining borders:

Measures to reduce *air pollution*

- reduce air pollution by tackling emissions, primarily generated by industry and domestic combustion
- cooperation on desulphuration of furnaces in the region, through technology transfer for instance
- retrofit lignite power stations in the Czech Republic, e.g. Prunérov I near Chomutov, with flue gas and effluent treatment plant
- set up an air monitoring network akin to the system practised in Bavaria (LÜB) in Vogtland in Saxony and in north-western Bohemia:
  - coordination with the Bavarian air monitoring system
  - regional availability of data
  - adherence to EU limits

Planning an interconnecting *landscape scheme*

(MODEL PROJECT F 1) on the basis of identified sensitive areas and involving the integration and demarcation of

- interconnecting biotopes,
- zones of "gentle" tourism,
- recreational amenities/green belt land serving to define development axes,
- rehabilitation schemes for devastated areas and contaminated surface water in

agricultural areas.

Measures to contain *woodland blight*:

- region-wide survey of extent of blight on a unitary basis,
- joint rehabilitation measures for forestry stocks (restructuring, fertilisation, liming)
- transnational coordination of measures (via forestry commissioners for instance).

Elaboration and implementation of joint transnational *species assistance programmes* (e.g. for the grouse, lynx, otter or black stork)

Establishment of a joint transnational *biotope monitoring project*

Drawing up of joint *waste disposal schemes* and concrete solutions inclusive of implementation thereof

Measures to improve the condition of waters:

- setting up of a region-wide system for monitoring pollution of surface waters

A total of 25 measurement packages for transnational cooperation in the field of ecology and environmental protection were agreed, thus fully taking account of the immense task in hand.

The "Planning an interconnecting landscape scheme" measure was picked out as a model project by the countries involved and registered for funding as a priority project to be implemented forthwith within the framework of the Interreg II programme. Owing to the project's interdisciplinary objectives, however, direct effectuation does not square with Interreg II's funding objectives and hence it is still awaiting implementation three years after the development plan was adopted. Given the variety of constituent aims in the project, in the first instance a feasibility study is being conducted.

Responsibility for initiating and conducting this "interconnecting landscape" case study lies with regional planning and is an element of its new remit, though in this case there is to be coordination with the Euregio Egrensis directorate. The feasibility study it has been determined must precede implementation of the project embraces the following tasks:

- scrutiny and coordination of measures with the aim of protecting, maintaining, and developing nature and landscapes as well as development of environmentally acceptable tourism and appropriate PR work and further training
- elucidation of funding modalities and options for co-financing
- clarification of executive basis of project
- indication of timetable for implementation together with funding schedule
- delineation of contextual relationship to overall scheme

The conception of nature and landscape conservation employed draws on Pfadenhauer (1991), envisaging a long-term strategy of developing an extensively balanced landscape, materially speaking, over the entire area into which a system of diverse modes of use can be integrated. This integrated implementation of the conservation objective is rooted in the experience that the preservation or "museumish" perpetuation of surviving residual biotopes is not enough since, instead of extending viable areas or where feasible linking them up, it merely leaves them as isolated patches. In the case of a large number of agrarian landscapes, by contrast, renaturation could facilitate protection of species on worked land, whilst changes in the intensity and nature of cultivation could minimise the need for land required as a buffer to contamination of soil and water. In respect of land use objectives comparable activities are discernible in

- the cultivation of drinking-water conservation zones and catchment areas for reservoirs by the water industry,
- the identification of natural woodland cells and/or especially protected and protected woodland by the forestry industry,
- the carrying out of water border strip programmes and the setting up of new retention zones for flood protection by the water industry,
- agricultural extensification and setting aside of land,
- the recultivation of landscapes exhausted by mining and deforested areas and, finally,
- plans to open up areas to tourism in conjunction with municipalities, tourist associations, etc.

This steady refinement of cooperation is an ideal means of taking the objectives of nature and landscape conservation further forward in a spirit of consensus without the need for compensation schemes. Alongside the daily routine of practical application, the inclusion of conservation, maintenance, and development objectives in regional planning and zoning represents the most effective lever for ensuring integrated conservation of physical landscapes. The process of drawing up regional plans and zoning modalities offers an opportunity to establish a broader base for the formulation and implementation of objectives in concert with other public interest bodies such as the main land users, the forestry commissions, and the departments for rural reorganisation. Agreement on common goals between the various interests involved urgently needs to be augmented by a bundling of funding programmes. The targeted application of funding currently relates to setting in train the most central stages towards ecologically sound land use as well as to tackling the most urgent issues together.

The course of implementation of the interconnecting landscape model project as a case study for a border region projects shows just how much separates the decision to adopt a common approach to border region development via the construction of mutually agreeable guiding principles and development goals from final implementation of same. This is true of the Euregio Egrensis and of the other, specifically Bavarian-Czech, border regions in which the disciplines of spatial planning and regional development, having initiated elaboration of development concepts, are wholeheartedly applying themselves to implementing the actual projects. Implementation and coordination at grass roots level are considered here to be the remit of regional development as part of the drive to test and refine new spatial planning tools.

Some five years have elapsed since a start was made on setting up border area concepts at the EU's external borders: the Trilateral Development Concept was elaborated in 1991 and adopted in 1994; border region development concepts for the Elbe/Labe, Ore Mountains/Krušnohori, and Neisse/Nisa regions were elaborated from 1992 to 1994 and the Spatial Development Master Plans for the German-Polish border region from 1992 to 1995. Transnational cooperation has additionally gained a significant impetus from targeted project support within the framework of the EU's Interreg II Community Initiative, in force since 1993. With so much having been done, it needs to be asked what cooperative planning and action have actually led to. This question acquires added poignancy given the present need for scientific and political debate on continuation of the Interreg programme and on border region funding in general (Myller 1997).When answering the question of what transnational cooperation has achieved, the yardstick of the "need for action" in border region development needs to be applied and the drawing up of specialist development objectives and their implementation scrutinised. The debate is grounded on sound documentary analysis, written questionnaires, interviews with experts and the evaluation of case studies from representative European border regions by the *Institut für ökologische Raumentwicklung* (IÖR), though emphasis in the latter instance is on the EU's eastern border. A systematic presentation of research findings from European border regions will form the basis of a separate IÖR publication.

3 Implementation of Development Concepts

3.1 Transnational Development Policies

Transnational development concepts for the EU's eastern border were elaborated in the early Nineties at the behest of regional planning bodies and regional policymakers in the Czech Ministry for Economic Affairs and also on behalf of the German-Polish Regional Planning Commission. From 1993 onwards, the relevant ministries of trade and industry commissioned the elaboration of plans of action and operational programmes. Each of the neighbouring countries commissioned their own experts to elaborate border region concepts. It is possible to make the following fundamental statements regarding the elaboration process:

1. Agreement was reached between the bordering countries concerning the drawing up of a joint development concept.
2. Interest was shown at national, regional, and municipal level in the findings of the strength/weakness analysis.
3. National representatives had a hand in formulating what action was needed in that more attention was given to the divergent conditions obtaining at the out-

set in the various countries when the modalities of the envisaged joint development process were thrashed out.

4. The guiding principles and goals of border region development were agreed unanimously and given a full airing at demonstration events in the various sections of the region.
5. When the catalogue of measures was compiled, it became evident that items where joint action was deemed necessary are weighted differently for either political or economic reasons, or both, by the various parties to the process (cf. 3).

The imbalance in the development concepts between items requiring action and measures that are mutually agreeable, an imbalance that made itself felt both quantitatively and qualitatively, stems from the causes listed.

From an assessor's point of view, the following closely intermeshed priority areas for the eastern EU border present themselves (listed in random sequence):

- stabilisation of demographic trends, most notably in the German sections, curbing of fluctuation with its negative consequences on age and qualification structures
- improvement of the technical infrastructure through the renewal of over-age transmission lines and plant and the building of roads, bridges, border crossings and paths into the neighbouring area where needed
- rehabilitation of environmental damage
- improvement of the accessibility of peripheral areas

The Euroregions regard the following priority areas of action in transnational regional development as long-term goals:

Tasks of transnational regional development from the Euroregion point of view

German view	Czech view
Economic development	Transport
Settlement	Technical infrastructure
Culture	Environment
Science	Economic development
Tourism	Agriculture

In keeping with the Interreg II programme (Ziegler 1997), the current range of activities of the Euroregions is directed towards putting the infrastructure in place as an important precondition for economic and demographic development as well as towards building up contacts at municipal level. The latter has a major impact on the building of trust and on the emergence of a sense of identity with the common region and is hence a basis for a targeted common mode of procedure.

Regarding the current state of development concepts, the formulation of objectives, and endeavours to get to grips with exactly what action is needed, the following *shortcomings* need to be noted:
Owing to the hierarchical system in place, responsibility for drawing up transnational development concepts is in the hands of ministers from the relevant departments and nations. Regional representatives are involved but are not directly tied by signature to implementation of the development objectives.

Under the aegis of spatial planning bodies, interdisciplinary development goals and measures for the border region are formulated in the development concept that require supradisciplinary shaping and coordination. The fact that the trade and industry departments are responsible for drawing up the plan of action and the operational programme causes the supradisciplinary view of the range of tasks to be focused solely on the one aspect, thus retrospectively downgrading the influence and scope for shaping the process enjoyed by spatial planning. This discipline is only an informal participant in the drawing up of plans of action.

The leading role of the trade departments also holds for the application for project funding, though the other affected bodies do take part and spatial planning is consulted.

In the case of the spatial planning development concepts, interdisciplinary groups of tasks have frequently been categorised as enjoying priority that, owing to the fact that departmental responsibility has been drawn too tightly, cannot be implemented within an interdisciplinary framework (cf. the case study "Interconnecting landscapes as they overlap with environmental protection, tourism, and agriculture") but have instead to be divided up into several project elements. Each of these is then required to comply with the applicable funding directive, and in the end the parts are no longer capable of contributing fully to the supradisciplinary development objective.

The efforts of the Euroregions to achieve interdepartmental and intercommunal cooperation are proving extremely laborious and, given the absence of executive powers for transnational regional development, they are often in vain where they are not underpinned by national development initiatives.

3.2 Implementation of Specialist Objectives

It is evident from the observations on the range of activities of the Euroregions in 3.1 that current undertakings in the border regions are being very strongly weighted according to their fundability and that the focus has been very much on finding suitable programmes for urgent cases, with a systematic tackling of all cases requiring action viewed as a more distant objective. This approach has given rise to a great number of successful project implementations and allowed human contacts to evolve, and this is very encouraging. Nevertheless, there is a danger that, by atomising activities, the overall objective will be lost from sight and that the hoped-for synergetic and pump-priming potential inherent in the measures given priority will be squandered and hence in the final instance inefficient use made of both EU and one's own funds.

As an aid to the further honing of transnational cooperation in the implementation of specialist objectives in the spirit of a systematic tackling of the key areas of action, reference may be made to the following *reasons for success in implementing border region projects*:

1. high personal input of players with specialist skills, powers of persuasion, and staying power;
2. good bilateral contacts and backing from private enterprise:
3. direct responsibility of a state department for project implementation or else informal consultation between ministries with access to funding;
4. high political interest at national level and high public interest in the region, e.g. involvement of the district authorities, of mayors, and of the media in these exercises;
5. link-ups between those in office and project managers as well as direct cooperation between departments at rapporteur level;
6. presence on European committees and mastery of the funding modalities;
7. awareness of the Euroregion and individual players on the part of the funding institutions.

Barriers to project implementation are shown to be:

1. differing sets of interests and items requiring action of the national neighbours, determined by differing points of departure;
2. differing, often changing powers of authority for border region work at national level; in Germany, overburdening of the administrative apparatus for approval of applications;
3. inadequate resources for co-financing from the region, too many sources of funding with prescribed funding objectives, absence of a funding programme for interdisciplinary objectives;
4. political imponderables concerning the observance of finance regulations, e.g. for the operation of sewage treatment plant on neighbouring territory;
5. insufficient expertise for specialist coordination and management tasks and inadequate knowledge of funding modalities;
6. local self-interest;
7. apprehension about interacting, plus political reservations, especially in questions of economic development and maintenance of the cultural heritage;
8. lack of attractiveness of project vehicle;
9. little updating of development goals and of the scientific material accompanying development tasks.

Since development concepts were put in place, the main areas in need of action that have been successfully tackled relate to the technical and social infrastructure, objectives that concur with the Interreg II remit. Success in promoting the objective of, and taking necessary action for, "economic development" is thin on the ground by contrast, confining itself to items such as the development of information systems, cooperation between chambers of industry and commerce on training measures, or joint presentation events and sales exhibitions. Also less than satisfactory is the processing of items requiring action in the structural sphere of ecology and environmental protection, which takes a bit of a back seat in the Interreg II programme. Likewise, the brief to stabilise demographic structures is making faltering progress via image concepts and training measures.

4 Recommendations from a Specialist Viewpoint on the Further Development of Border Regions

Studies on the degree of implementation of specialised development objectives and border region projects underline the correctness and necessity of jointly drawing up interdisciplinary development concepts so as to be able to accurately pinpoint areas requiring action and use them as the basis for strategic advances in a process of phased development. When tackling items requiring action, *a dual mode of procedure* is advised.

Firstly, formulation of medium to long-term development targets within the framework of an envisaged overall goal and, *secondly*, border region projects assigned to this vertical target orientation as horizontal planes and classified within the overall target concept according to urgency of implementation, pump-priming potential, feasibility and political efficacy or needs so as to constitute a tiered programme of immediate action.

The tendency to tailor activities to available funding programmes needs increasingly to make way for a target-oriented approach and specialist preparation, at the same time eradicating the causes of such knee-jerk responses. The following recommendations can be drawn from the studies:

1. Funding programmes need to be more flexibly designed and to be geared towards areas of priority action. In the process, specialist goals need to be integrated, not superimposed or bracketed out.
2. Border region development concepts need to be drawn up at interdisciplinary level under the aegis of the regions and with the participation of specialist departments at national level. The same should apply with regard to the responsibility of a region for updating a concept and to research accompaniment when projects are put to effect or in the course of coordinated tasks.
3. The principle of sustainability needs to be accorded due attention both when concepts are drawn up and when coordinating implementation of projects, thus furnishing proof of the pioneering role of the border regions in the process of European integration and reorganisation.
4. Effectuation of development concepts in the form of a flexible shaping of the overall concept should also be the responsibility of the border regions. This

task implies interdisciplinary coordination of areas of activity. The specialist input of the disciplines involved needs to be made practical use of to this end.
5. Border regions need to be vested with regulatory powers for the implementation of the target concept and projects including the requisite supradisciplinary coordination. Setting up a Regional Fund specifically for efficient project implementation is advisable.
6. The present definition of border regions needs to be reviewed taking functional interdependences into account and the areas impacted upon by the border redefined. The affected areas should share exclusive responsibility for development processes in the border regions and be included in regional self-commitment.
7. Representatives of the border regions should be elected in the region and be given a say in regional planning and national funding policies. They should be directly represented in the EU policy-making process.

References

Bundesministerium für Raumordnung, Bauwesen und Städtebau (1995) Raumordnerische Leitbilder für den Raum entlang der deutsch-polnischen Grenze, Bonn

Malchus V von (1994) Erfahrungen der grenzübergreifenden europäischen Zusammenarbeit im Rahmen raumordnerischer Orientierungen. In: Roch I (ed) Umweltverträgliche Regionalentwicklung bei besonderer Berücksichtigung der Grenzräume. Dresden: Institut für ökologische Raumentwicklung, IÖR-SCHRIFTEN, vol 9, pp 77–83

Moll P (1995) Kooperation an der EU-Binnengrenze – Das Beispiel des SaarLorLux-Raumes. In: Roch I (ed) Umweltverträgliche Regionalentwicklung bei besonderer Berücksichtigung der Grenzräume. Dresden: Institut für ökologische Raumentwicklung, IÖR-SCHRIFTEN, vol 9, pp 71–77

Myller R (1997) Entwurf eines Berichts über grenzüberschreitende Zusammenarbeit. Regional Policy Committee, European Parliament

Pfadenhauer J (1991) Integrierter Naturschutz. In: Garten + Landschaft, vol 1/1991, pp 13–17

Roch I (1995) Die gesamteuropäische Bedeutung der deutsch-polnischen Grenze. In: Regionen im Umbruch/Regiones en transición, REGIO No 6, pp 175–185; Roch I (1995) El significado de la frontera germano-polaca para europa. In: Regiones en transición, compilador. Alejandre J A (ed) Universidas de Guadalajara, Guadalajara pp 309–325

Roch I et al. (1995) Trilaterales Entwicklungskonzept für den bayerisch-sächsisch-tschechischen Grenzraum. In: Jurczek P (ed) Beiträge zur Kommunal- und Regionalentwicklung, No 14

Roch I, Leimbrock H et al. (1996) Das Sächsische Vogtland im Dreiländereck Bayern-Böhmen-Sachsen. Ansätze einer umweltverträglichen grenzübergreifenden Entwicklung. Dresden: Institut für ökologische Raumentwicklung

Scott J (1997) Empfehlungen für ein grenzübergreifendes Regionalmanagement, unpublished manuscript, Berlin

Selke W, May R (1994) Ansätze und Ziele einer europäischen Raumordnungspolitik. In: Roch I (ed) Umweltverträgliche Regionalentwicklung bei besonderer Berücksichtigung der Grenzräume. Dresden: Institut für ökologische Raumentwicklung, IÖR-SCHRIFTEN, vol 9, pp 83–86

Ziegler A (1997) Empfehlungen für die grenzübergreifende Regionalentwicklung, Finanzpolitische Aspekte, unpublished manuscript, Düsseldorf

Cross-Border Development in the Vienna/Bratislava Region: A Review[1]

Wilfried Altzinger and Gunther Maier
Vienna University of Economics and Business Administration
Vienna, Austria

Jarko Fidrmuc
Institute for Advanced Studies
Vienna, Austria

1 Introduction

The economic and political changes in central and eastern Europe underway since the late 1980s have fundamentally changed the European landscape and the framework for economic activity in major parts of the continent. At a time when the European Union was moving towards a higher level of integration and aiming for a free flow of labour, investment, and goods, the opening of the former Eastern Bloc provided new and additional opportunities whilst posing new challenges.

More or less overnight, a huge new market with enormous reserves of demand opened up whose future prospects to a large extent depend on the success of processes of restructuring and privatisation in the former socialist countries. The restructuring process has been accompanied by high levels of uncertainty, decline in production, unemployment, and social degradation. In their struggle to find new markets as a substitute for CMEA trade which collapsed early in the transition, most countries turned towards the west and tried to penetrate its abundant markets. There followed a process of mutual market penetration that provided golden opportunities (at high risk) for some sectors, and fierce new competition for others. In the west, too, this process of integration with central and eastern Europe (CEE) triggered a fundamental and quite massive restructuring process.

The labour market plays a central role in such processes. The low level of wages in the post-socialist countries is one factor attracting investment from labour intensive industries, thus threatening jobs and wage levels in the west (for a more detailed discussion see section 3.2). At a time of high and constantly increasing levels of unemployment, the integration with CEE adds to friction in the labour market and social system. This friction is intensified by workers from CEE looking for higher

[1] An earlier version of this paper was presented at the 36th European Congress of the Regional Science Association, ETH Zürich, Switzerland, 26–30 August 1996.

paid jobs in western countries. The impact of this on the labour market (massive loss of jobs in some sectors and/or regions) and corresponding fears in the population have led to restrictive migration and labour market access policies in some countries. As noted by OECD in its most recent migration report, this policy is short-sighted because it does not take into account the relationship between employment, trade, and foreign direct investment: "*International cooperation to control flows is only a partial response to the intensification of migration movements. It is for this reason that the idea of sustainable development as a means of eventually reducing the incentive to emigrate in countries with high emigration potential is gaining ground among OECD countries.*" (OECD, 1995a; p 52).

Nowhere are the problems and complex interdependencies of restructuring and integration more clearly visible than along the borders between today's EU and the former socialist countries of central and eastern Europe. The area where agglomerations of population and economic activity on both sides of the border are in closest proximity is the Danube valley, with Vienna, the capital of Austria, on one side of the border, and Bratislava, the capital of Slovakia, on the other. With only 55 kilometres from centre to centre, the two cities are one of the closest located pairs of capitals in the world. Without the history of economic separation and the current development differential this has given rise to, Vienna and Bratislava today would probably form an integrated agglomeration, and a major economic force in the centre of Europe to boot. Because of this, we shall refer to the two cities as the "Vienna/Bratislava region". However, one of the aims of this paper is to investigate to which extent this sobriquet is justified.

Because of the proximity of the two cities and their high level of economic activity, it follows that cross-border development processes generated by the revolution in eastern Europe are particularly pronounced in this region. In this paper we try to review these processes.

Although it is the labour market of the region that specifically interests us, we shall also be looking at the development process in a broader context. Given the interdependencies alluded to above, we argue that labour market processes in this region can only be properly understood if they are viewed in conjunction with trade and capital mobility. This approach is strongly supported by theoretical arguments, but nevertheless often neglected in favour of a more sectoral approach.

Cross-border development in the Vienna/Bratislava region is of great significance for a number of reasons. For the region itself, development over the next decade on both sides of the border will be crucial to its longer-term prospects. The future of a population of more than three million will be decided within this region. For policy makers at the European level, the Vienna/Bratislava region is something of a test case for the tasks and developments they will face when they integrate Slovakia and other post-socialist countries into the Union. The success or failure of this micro-scenario may well enhance or diminish the very likelihood of expansion eastwards. For regional economists the Vienna/Bratislava region represents a rare example of a laboratory-like setting. Where else can they study what happens when a proximity is established between two large cities that – in economic terms – have for decades been separated?

In the following section we shall set forth the theoretical background to the paper in more detail. To this end, we shall be discussing the relationships between labour migration, capital migration, and trade flows. Section 3 then turns to the empirical evidence. In three subsections we shall discuss trade (3.1), direct investment (3.2), and labour market related developments (3.3). The paper ends with a concluding section.

2 Theoretical Background

The main theoretical question is, what economic mechanisms will come into being when two regions with differing levels of development are located next to each other within a unified economic framework, and what effect will these mechanisms have on the development levels of each. In short, we are intent on defining the development trajectories of the two regions.

The standard answer to this question is provided by the Heckscher-Ohlin-Samuelson model (e.g. Heckscher, Ohlin, 1991; Mundell, 1957) of neoclassical economics. For two regions with identical production technologies but different factor endowments it predicts converging factor prices.

In an integrated system of regions, this is the result of two economic forces[2]: trade and factor mobility. With free interregional trade each region will specialise in those products in which it has a comparative advantage over the other. These are products whose manufacture presupposes the factor that is relatively abundant in the region. In the form of products traded, each region exports the more abundant factor of production and imports the scarcer one in this way. This, it will be noted, implies that the two regions will be exporting different products. Through this mechanism, free trade will balance factor price differentials between the regions and work towards further convergence of development paths. When factors are mobile between the regions, the difference in factor prices arising from the difference in endowments will stimulate migration of labour and capital. Each factor will flow from the region where it is relatively abundant – and cheap – to the one where it is scarce – and expensive. Since factors are compensated by their marginal products, factor mobility tends to equilibrate factor prices between the regions.

Three important results can be posited using the Heckscher-Ohlin-Samuelson model:

1. Capital and labour will flow in opposite directions;

[2] This is the case even when no goods or factors are exchanged between the regions, simply because of capital accumulation: With identical savings and depreciation rates the region with lower capital intensity will accumulate capital faster and thus grow faster than the more capital intensive region.

2. The regions will specialise in different products;
3. Trade and factor mobility are substitutes.

This model establishes a close (substitutive) relationship between direct investment (capital mobility), migration (mobility of labour) and trade. This has an important policy implication: when one of these mechanisms is constrained, the others will be stimulated to work towards equalising factor price differentials. Or, to put the same argument in a more positive way: by liberalising one of these areas, policy makers can relieve economic pressure in another. For example, liberalising trade and capital flows should help to equalise the main economic push and pull factors which encourage migration, namely wage differences and uneven economic development.

However, the price equalisation theorem in the Heckscher-Ohlin-Samuelson model rests on some very restrictive assumptions. One of these assumptions is that the two regions use identical – or very similar – production technologies. This assumption cannot hold for the specific East/West dichotomy in Europe. It needs to be noted that differences in production technologies have not only technical but also institutional, organisational and administrative dimensions. The latter factors seem to be particularly important in the context of eastern Europe. "Much of the blame must rest on the Communist legacy of distorted incentives, bad management, and low morale" (Layard et al., 1992, p 56). Layard et al. (1992) show in a very simple model that, given (sufficiently large) efficiency differences, trade alone cannot equalise factor prices. Even when factor prices converge, this process will come to a halt before parity is achieved. When efficiency differences are large enough, labour and capital may even flow in the same direction because prices for both factors are higher in the better-developed economy. The only solution for such a situation is to improve efficiency at both the micro and the macro level. Besides institutional and administrative reforms, thoroughgoing structural reform at enterprise level can also serve to achieve this. Intensive trade relations will aid the adoption of foreign management routines and, by example, improvements in efficiency at the micro-level. In turn this will attract new foreign investment and stimulate economic growth.

Furthermore, the second result listed above is in stark contrast with the common empirical evidence of intra-industry trade. The lion's share of world trade occurs between similarly structured industrialised countries and within certain industries rather than between the industrial and the non-industrialised world. The exports and imports of developed countries show a similar composition, an observation that is hardly compatible with the specialisation argument put forward in the Heckscher-Ohlin-Samuelson model.

In the literature on regional economics, alternative arguments have flourished since the 1950s. Myrdal (1957) and Hirschman (1958) were the first to argue that the mechanisms of the Heckscher-Ohlin-Samuelson model may actually lead to divergent rather than convergent development paths (see e.g., Maier, Tödtling, 1996). They point out that factor mobility is always highly selective and that modern economies are dominated by monopolies, oligopolies, and scale economies, structures that are incompatible with the neoclassical world of perfect competition. As a result, free trade and factor mobility may trigger a circular process of cumulative causation that will tend to deepen development differences rather than eliminate them.

Unfortunately, this polarisation argument has been largely ignored by standard economic theory. One of the reasons for this is that, since it has never been formalised, the polarisation argument is not capable of being turned into a consistent set of statements that could compete with the neoclassical theory.

Only recently have these arguments entered the mainstream of economic literature. In an attempt to endogenise technical progress[3], Romer, Grossman and others (see, e.g., Romer, 1986, 1990; Grossman, Helpman, 1992; Bröcker, 1994) arrived at similar conclusions as the representatives of polarisation theory thirty years before: externalities or not perfectly competitive markets are necessary conditions for the production of research and development. These conditions also produce increasing returns to scale and thus cumulative feedback mechanisms that weaken or even eliminate the equilibrating tendency of the Heckscher-Ohlin-Samuelson model. Krugman (1991), for example, demonstrates that because of scale and agglomeration economies, some random event may put otherwise identical regions on different and divergent growth paths. The question as to whether free trade and factor mobility will eliminate interregional differences becomes an empirical rather than a theoretical one. Barro (1991) and Barro and Sala-i-Martin (1991), for example, try to answer this question and find evidence for convergence. However, according to their results it will be about 35 years before half the development differential is eliminated; that compares to 12 years in a neoclassical model.

Although these findings raise serious doubts about free trade and factor mobility as mechanisms for eliminating development differentials, one cannot elicit from them reasons for hampering these mechanisms. They are necessary, but not sufficient conditions for the elimination of development gaps. For thc given East-West scenario in Europe, this suggests that one should be trying with the aid of institutional reform, stable and predictable economic policy, reliable infrastructure etc. to improve productivity at the micro as well as at the macro level.

International economic cooperation, free trade, and foreign direct investment all provide important incentives in this direction.

---

[3] Technical progress is an exogenous factor in the standard neoclassical model. This is highly unsatisfactory since in the long run economic growth is driven only by technical progress.

3 Empirical Evidence

In this section we shall be reviewing the empirical evidence as regards development transfer mechanisms in the Vienna/Bratislava region. We investigate trade, foreign direct investment, and the labour market in separate subsections. Not all information for the regional level is available, however. We shall accordingly need to combine evidence on different spatial scales. This applies particularly to trade information available only at national level, also bearing in mind that, for the first years, the one nation involved was still Czechoslovakia.

Despite the fact that Austria and Slovakia were divided by the Iron Curtain for over forty years, they share a long common history (Maier, Mašek, 1992). This made it easier for the two countries to commence common initiatives after the collapse of Communism. Another important factor is the close proximity between their capitals, Vienna and Bratislava, which in turn makes both measuring and controlling development transfer mechanisms somewhat difficult. For example, since Bratislava is within easy commuting distance of Vienna, labour mobility is not confined to migration. A substantial amount of Slovak labour in Austria may be commuters who will not show up in the migration statistics and will not be confined by administrative constraints on labour migration. A similar argument can be made with respect to foreign trade, which will be augmented by cross-border shopping trips in the region. Such factors are hard to quantify but need to be borne in mind in the following analysis.

3.1 Austria's Trade with Central and Eastern European Countries (CEEC) and notably with Slovakia, 1988 to 1996

The opening up of the Iron Curtain provided significant momentum for Austria's foreign trade with CEE countries:

1. The emergence of a huge new potential market (very much depending on the way these economies develop). The complete breakdown of CMEA trade has quite substantially aided this development.
2. Intensified competition in several, primarily labour, energy and commodity-intensive sectors.
3. With direct CEE investment by foreign companies up, growing levels of intra-company trade.

We should nevertheless not lose sight of the fact that turning this potential to effect will require (complete) trade liberalisation.

Table 1 illustrates developments in Austrian trade with the CEEC between 1989 and 1996.[4] Austria has benefited enormously from the fall of the Iron Cur-

---

[4] We use estimates of Austrian trade for 1996 provided by the Austrian Chamber of Commerce (BWK, 1997).

tain, both as regards trade volume and the trade balance. Austrian exports to the CEEC (Hungary, Poland, Slovakia, Slovenia and the Czech Republic) rose by almost 240 % between 1988 and 1996. Austrian exports to what was Czechoslovakia actually increased cumulatively by 375 % in the relevant period. In turn, Austrian imports from the CEEC increased by 150 % (290 % in the case of what was Czechoslovakia) since 1988.

Table 1. Austrian Foreign Trade with the CEEC

Growth of Exports, %	1989	1990	1991	1992	1993	1994	1995	1996[a]
former CSSR/CSFR	6.83	72.51	6.04	50.74	11.74	16.08	19.85	3.62
Czech Republic						18.35	17.74	-1.24
Slovakia						9.79	26.15	17.14
Poland	40.74	-16.51	70.89	-5.55	-8.84	-6.42	30.26	15.53
Hungary	27.13	20.75	38.66	7.09	6.37	21.15	4.84	-3.71
former Yugoslavia	18.16	34.95	-23.00	-2.54	11.63	28.75	21.84	-1.21
Slovenia					20.69	17.49	22.11	-5.07
FSU	4.09	-12.18	-7.35	-13.95	-5.19	21.03	20.81	-2.31
CEEC	24.21	24.14	32.66	16.90	5.46	14.49	14.44	2.42
Eastern Europe	14.05	14.73	8.42	7.23	4.86	17.46	17.96	1.13
Total exports	12.03	8.56	2.78	1.78	-4.18	9.71	13.17	5.53
Growth of Imports, %	**1989**	**1990**	**1991**	**1992**	**1993**	**1994**	**1995**	**1996[a]**
former CSSR/CSFR	11.34	-4.86	16.06	48.96	10.80	25.60	16.06	14.60
Czech Republic						23.64	11.72	5.68
Slovakia						31.28	27.95	35.93
Poland	2.67	15.18	12.82	-11.37	-6.62	10.40	23.05	3.86
Hungary	23.11	11.44	31.41	4.17	-9.47	18.51	-1.97	14.74
former Yugoslavia	28.19	7.14	-9.16	-12.12	-4.03	17.03	28.55	1.97
Slovenia					12.65	20.26	29.61	2.92
FSU	-1.28	20.18	-4.79	-11.38	1.39	35.03	2.60	-6.35
CEEC	13.63	6.50	21.91	14.15	-0.96	20.28	10.22	12.79
Eastern Europe	11.36	9.07	9.23	4.61	-1.39	23.57	10.30	7.68
Total exports	14.01	8.07	6.41	0.34	-4.89	11.32	6.23	5.59
Balance, ATS billion	**1989**	**1990**	**1991**	**1992**	**1993**	**1994**	**1995**	**1996[a]**
former CSSR/CSFR	-1.725	2.235	1.729	2.738	3.163	2.503	3.584	1.748
Czech Republic					2.213	2.136	3.196	2.283
Slovakia					0.949	0.367	0.388	-0.537
Poland	0.888	-0.638	1.820	2.048	1.755	0.856	1.487	2.459
Hungary	0.837	1.741	3.047	3.599	5.724	7.219	8.443	5.808
former Yugoslavia	3.200	5.987	3.721	8.280	5.476	7.629	8.908	8.110
Slovenia				5.642	3.434	3.941	4.508	3.859
FSU	11.473	10.076	9.335	8.033	-1.146	-2.614	-1.003	-0.527
CEEC	0.000	3.338	6.596	8.385	10.642	10.578	13.513	10.015
Eastern Europe	15.898	20.682	20.762	25.632	16.247	16.461	23.481	19.255
Total balance	-85.377	-90.167	-112.869	-106.367	-97.739	-116.363	-88.017	-93.273

[a] estimates of Austrian foreign trade for 1996 provided by the Austrian Chamber of Commerce (BWK) on the basis of statistics available in partner countries. *Source: own calculations.*

For trade between non-Soviet eastern Europe and the EU, Collins and Rodrik (1991) calculate that the 1988 volume is too low by a factor of 3.5 as compared to more liberalised regimes. Hamilton and Winters (1992) use gravity models that relate bilateral trade flow to aggregate output for both countries (proxies for potential supply and demand) and distance between capitals (proxy for transport and transaction costs) as a means of arriving at projections for trade between eastern European countries and selected OECD countries (excluding Austria). Holzmann and Zukowska-Gagelmann (1996) provide similar trade projections for a larger number of countries. A comparison between these projections and trade realised reveals that Austrian trade is still significantly below capacity. The findings suggest that Austrian exports to the CEEC (including Slovenia and Croatia) account for about 30 % of trade potential, while imports from these countries attain only 20 %. However, this may be due in large part to higher discrepancies in the cases of Croatia and Poland. Trade with direct neighbours seems to be much closer to the potential mooted (see Fidrmuc, 1997).

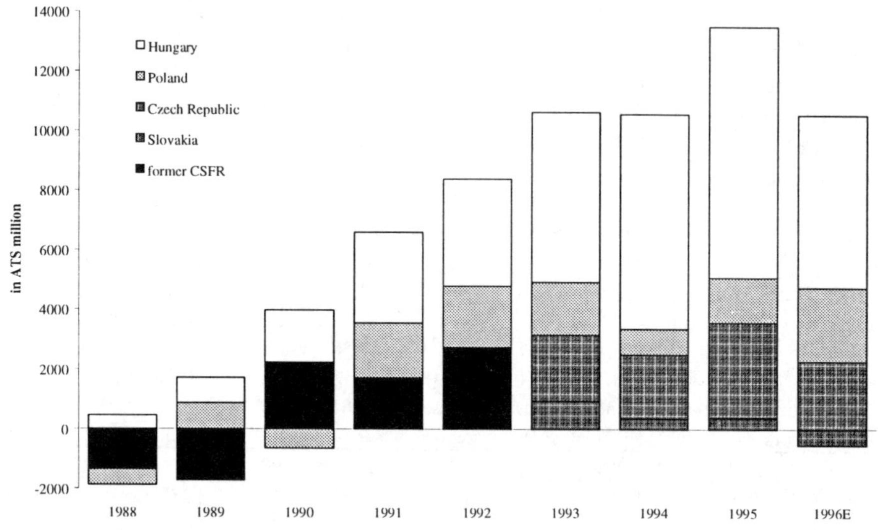

Source: Own calculations.

Figure 1. Austria's Trade Balance with the CEEC

Austria's trade balance with CEE countries improved from 1.4 billion ATS in 1988 to 13.5 billion ATS in 1995 (Austria's overall trade balance was minus 88 billion ATS in 1995 having stood at minus 116 billion ATS in 1994!). This is mainly due to the huge trade deficits of Hungary and former Yugoslavia. But Austria's trade balance with the former CSSR/CSFR likewise improved from a deficit of minus 1.4 billion ATS in 1988 to a surplus of 3.6 billion ATS in 1995. Austria's balance with the former CSFR worsened in 1994 for the first time in this period because of growing exports from the Slovak Republic to Austria, which increased by 33 % in this one year (Austrian exports to Slovakia grew by only 10 %). Thus Slovakia's trade deficit with Austria declined rather sharply in 1994 and for 1996 is actually estimated to have been 'negative'. This is mainly the result of economic recovery in the EU and further trade liberalisation.

However, a trade surplus with CEE countries that helps Austria reduce its overall trade deficit contributes to severe problems for some countries in eastern Europe. They will not be able run up such large deficits in the long term and will have to take measures to achieve more balanced foreign trade figures. The Czech Republic recently set the scene by raising a 20 % levy on import prices. We may also be seeing similar measures from other CEE countries before long.

We see manufactured imports constituting a growing share of overall Austrian imports from the CEEC (see Figure 2). The share of manufactured imports from the four CEE countries increased from only 30 % in 1988 to about 60 % in 1995. Manufactured imports as a share of overall Austrian imports from Slovakia were comparable to those involving the other CEEC, despite the higher volatility that is to be expected in trade with smaller countries. However, this share increased sharply to more than 70 % in 1995. Note that total Slovak exports came on very impressively in 1995. Nevertheless, these shares still remain significantly below the ratio of manufacturing to total goods exported by Austria to the CEEC, which remained at around 90 % for the whole of the period under review, despite some fluctuations at the end of the 1980s.

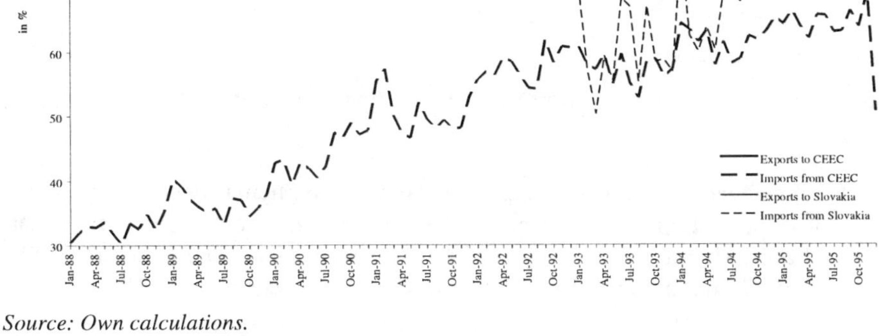

Source: Own calculations.

Figure 2. Share of Manufactured Exports and Imports in Austrian Trade with the CEEC

Moreover, the ratio of manufactured to total goods imported by Austria from Slovakia is lower than the ratio of manufactured to total goods exported by Slovakia to the EU (see Table 2). Although Slovakia has succeeded in shifting trade away from former CMEA markets towards more economically developed countries (mainly the EU), her breakdown of trade has remained relatively static. That means that regional losses in other eastern European countries and the former Soviet Union have been sufficiently compensated in OECD countries. The share of manufactured exports declined from 87 % in 1991 to 85 % in 1994 (for better comparison, we ignore Slovak trade with the Czech Republic for the period since 1993), but it was back up to 88 % in 1996. Similarly, the share of manufactured to total goods exported by Slovakia to the EU increased from 84 % in 1991 to 91 % in 1996.

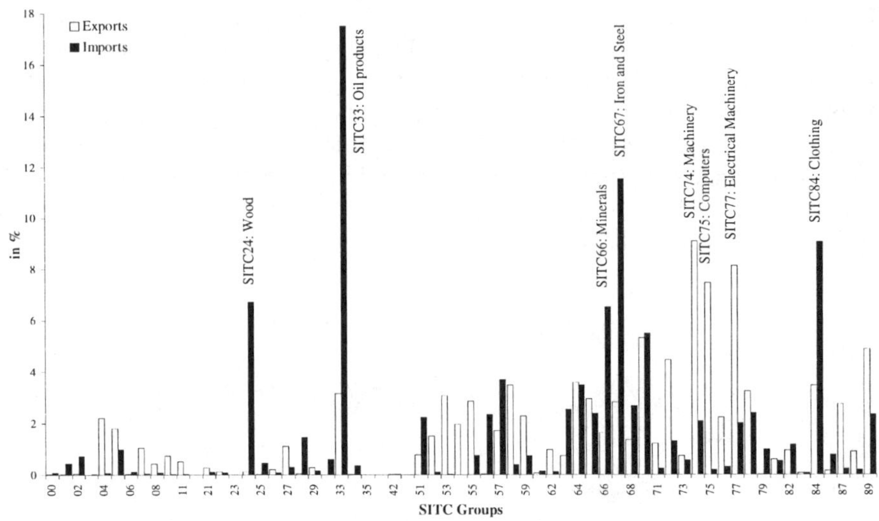

Source: Own calculations.

Figure 3. Structure of Austrian Exports to and Imports from Slovakia in 1994

The import side of Slovak foreign trade shows a particular dependence on imports of mineral fuels and raw materials from Russia. Nevertheless, this share declined quite sharply from 36 % in 1991 to 18 % in 1996. The structure of Slovak imports from Austria is very similar to that for imports from the EU.

Table 2. Commodity Structure of Slovak Foreign Trade by SITC Classification

Exports	foodstuffs	beverages	raw materials	mineral fuels	chemical products	semifinished goods	machinery	miscellaneous manufac. prod.	manufactured products	Total
Total, 1991	6.8	0.7	4.6	1.1	12.0	36.5	22.4	15.8	86.7	100.0
EU12, 1991	8.3	0.0	6.9	0.7	13.4	43.7	12.1	14.6	83.8	100.0
Total, 1994	4.5	0.9	5.1	4.6	12.8	39.2	19.4	13.3	84.7	100.0
EU12, 1994	2.6	0.0	4.4	0.1	11.3	42.8	22.4	16.2	92.7	100.0
Austria, 1993	5.0	0.0	12.6	17.8	13.5	29.8	12.5	8.7	64.5	100.0
Total, 1996	3.1	0.5	4.7	4.0	11.0	38.8	24.5	13.4	87.7	100.0
EU15, 1996	1.8	0.1	4.5	2.7	10.0	37.7	29.0	14.3	91.0	100.0
Imports	foodstuffs	beverages	raw materials	mineral fuels	chemical products	semifinished goods	machinery	miscellaneous manufac. prod.	manufactured products	Total
Total, 1991	4.3	0.8	11.8	36.1	8.8	8.1	24.3	5.8	47.0	100.0
EU12, 1991	5.8	1.1	5.4	1.2	17.9	9.2	51.4	7.9	86.4	100.0
Total, 1994	7.0	1.3	5.3	19.3	13.3	16.9	27.7	8.9	66.8	100.0
EU12, 1994	8.2	0.9	2.5	0.6	17.1	15.4	43.3	11.4	87.2	100.0
Austria, 1993	8.9	1.1	1.7	1.7	14.2	14.8	45.2	12.1	86.3	100.0
Total, 1996	5.7	0.4	5.7	19.8	11.1	10.8	37.3	9.0	68.2	100.0
EU15, 1996	6.2	0.6	3.4	1.3	15.3	14.8	48.6	10.0	88.7	100.0

Source: Fidrmuc, J. and Fidrmuc, F. (1995); own calculations.

Table 3. Role of Austrian FDI (Cumulated Figures) in Central Eastern Europe

Year	FDI Units/Total						FDI Units/Austria						Austrian Share (%)					
	CZ	SK	HU	SLO	PL	RO	CZ	SK	HU	SLO	PL	RO	CZ	SK	HU	SLO	PL	RO
1989	NA		157	218	NA	NA	NV		NA	NA	NA	NA	NV		NA	NA	NA	NA
1990	700		3,971	834	1,645	1,529	212		NA	NA	180	30	30		NA	NA	11	2
1991	5,500		9,613	1,325	4,796	7,897	1,300		3,507	269	322	215	24		36	20	7	3
1992	6,175	2,825	13,714	2,815	10,131	20,677	1,092	908	4,087	613	420	480	17	32	30	22	4	2
1993	14,052	5,143	18,000	3,169	15,053	29,134	2,450	1,141	4,700	670	500	664	17	22	26	21	3	2
1994	23,005	7,207	24,153	3,660	20,000	38,356	2,900	1,288	5,400	700	550	991	13	18	22	19	3	3
1995	32,946	8,740	27,900	4,007	24,000	42,464	3,240	1,380	5,550	NA	600	1,103	10	16	20	NA	3	3
June 1996	37,493	9,419	25,500	NA	26,332	43,951	3,300	1,412	5,500	NA	NA	1,135	9	15	22	NA	NA	3
Year	FDI Stock (USD million)						FDI Stock/Austria (USD million)						Austrian Share (%)					
	CZ	SK	HU	SLO	PL	RO	CZ	SK	HU	SLO	PL	RO	CZ	SK	HU	SLO	PL	RO
1989	NV		550	165	NA	NA	NA		53	NA	NA	NA	NA		10	NA	NA	NA
1990	670		1,450	501	NA	108	18		299	NA	NA	NA	3		21	NA	NA	NA
1991	1,314		3,064	827	700	264	104		406	165	26	NA	8		13	20	4	NA
1992	1,556		4,705	1,044	1,400	533	76		760	211	120	9	5		16	20	9	2
1993	2,053	366	7,186	1,113	3,100	761	126	90	958	198	195	17	6	24	13	18	6	2
1994	3,077	552	8,506	1,265	4,300	1,244	217	114	1,295	203	160	28	7	21	15	16	4	2
1995	5,797	740	12,700	1,643	6,832	1,595	316	160	2,150	277	248	39	5	22	17	17	4	2
June 1996	6,039	807	13,250	1,700	10,155	1,920	353	169	2,050	290	316	42	6	21	15	17	3	2

Source: Austrian Ministry for the Economy, 1997.

As can be seen from a comparison of Table 2 and Figure 2, Slovak trade with Austria deviates from this EU pattern. This is to some extent the result of intensive cooperation between the Slovnaft Bratislava and OMV Schwechat refineries (see Figure 3). In 1994, Austrian imports of fuels accounted for about 20 % of imports from Slovakia (with oil-based products making up 18 % of this), whereas fuels comprised less than 5 % of Slovak exports overall. The ratio of raw materials to total goods imported by Austria from Slovakia (10 % in 1994) is also higher than the mean for Slovak exports (5 % in 1994). This discrepancy primarily reflects specific cooperation ventures between Austrian and Slovak firms.

Although trade between Slovakia and the EU has been making satisfactory progress, some significant impediments remain. The most difficult structural problem for Slovak exports is the heavy concentration on *sensitive commodities* (Fidrmuc, J. and Fidrmuc, F., 1995; p 18). This is partially the result of the former intra-CMEA division of labour which located a fair amount of heavy industry and mining in Slovakia. Direct foreign restrictions (mainly tariffs and quotas) still represent a significant impediment to exports of sensitive commodities. An analysis of former CSFR exports in 1991 shows that 37 % of all exports are 'strongly sensitive commodities' (i.e. foodstuffs, steel and iron, motor cars) and 69 % 'sensitive commodities' (i.e., in addition to the products already mentioned, textiles, building materials, timber products, arms) (Fidrmuc, J. and Fidrmuc, J., 1995; p 47). Although the Associative Agreements with the EU, signed in 1991 with the former CSFR and amended for the Slovak Republic in 1995, provide for a gradual elimination of tariffs and quotas, some new ones were introduced in 1992.

Protectionist measures are nevertheless less prevalent in Bratislava due to its better industrial structure. According to the Slovak ministry for the economy, the most important exporting firms in Bratislava are Slovnaft Bratislava (refinery products, 4.68 % of Slovak exports in 1996) and Volkswagen Bratislava (motor cars, 4.08 % of exports). These firms rank directly behind the largest Slovak giant, the East Slovak Iron Works VSZ located in Kosice (iron and steel, 6.37 % of exports). In the region of Bratislava the list is augmented by Siemens Automotive Bratislava, which exports a number of consumer durables (1.22 % of exports).

From the Austrian point of view, products like cement, steel, textiles and clothes as well as agricultural machinery were a matter of some concern (Aiginger, 1995). “Quota regulations and anti-dumping measures by Austria were an attempt to gain time and adapt these branches to the new circumstances.” (Breuss and Schebeck, 1996; 100)

3.2 Foreign Direct Investment (FDI)

The main goal for most FDI is to improve the competitiveness of a multinational company at the micro level. In so doing, it may also enhance the general performance of both countries involved at the macro level. Whether this really happens or not will depend on specific circumstances, viz. the amount of capital transferred, the capacity to innovate and absorb new technologies, the extent to which organisational structures and management performance improve etc. (UNCTAD, 1995). A crucial factor in this context is the level of human capital in the host country.

Although it is somewhat difficult to measure the impact of FDI on economic growth, most studies conclude that FDI is conducive to technology transfer and contributes markedly to economic growth (i.e. Borensztein, et al., 1995). However, the situation is a little bit more complex when FDI is directed towards regions or sectors that have been characterised by strong monopolies until recently. This is the case for all CEE countries. Summarising such a situation, an ECE report concludes: "It is in countries where economic and social institutions are relatively weak, where the structure of the state is weak – that predatory behaviour by the enterprise is both more apparent and least restrained. The Transition countries are, by definition, characterised by still imperfectly functioning markets, embryonic institutions ... Where markets are not competitive or the state lacks effective regulatory powers, there can be no presumption that the activities of foreign companies will automatically contribute to a more efficient use of resources" (ECE, 1995; p 14).

In addition to this feature one should note that the most important reason for FDI in eastern Europe is market access rather than low labour costs. A survey of the EBRD concludes: "Most striking perhaps is the predominance of market access among factors of importance to investor decision ... Factor cost advantages are clearly rated as less important than market access in all the surveys ... Most studies explicitly play down the role of cheap labour" (EBRD, 1994, p 132). A survey of the Austrian National Bank shows the same results. For more than 85 % of all Austrian investors in Hungary and the Czech Republic market access was given as the main motive.

Table 4 shows the distribution of FDI in eastern Europe. It is not surprising that the major tranche (nearly two thirds) of all FDI is concentrated in Hungary, Poland and the Czech Republic. To measure the amount of FDI in relative terms it is useful to compare the stock of investment to GDP. This again shows the preeminence of Hungary as an investment destination (27.9 % in 1995) but even the Czech Republic shows remarkably high figures (12.3 %). Both Poland and the Slovak Republic lag behind on 5.7 % and 4.3 % respectively. Certainly, there are several reasons for this, e.g., lack of political and financial stability, inadequate legal institutions and frameworks and – in Slovakia – an inadequately implemented privatisation programme.

Table 4. FDI and investment stocks as percentage of GDP in central eastern Europe

	Cumulated FDI Stock (million USD)			FDI as percentage of GDP (%)		
	1994	1995	1996[a]	1994	1995	1996[a]
Czech Republic	3,077	5,797	6,039	8.55	12.28	11.52
Slovak Republic	0,552	0,740	0,807	4.00	4.25	4.29
Hungary	8,506	12,700	13,250	20.50	27.91	28.93
Slovenia	1,265	1,643	1,700	8.85	8.83	9.19
Poland	4,300	6,832	10,155	4.55	5.67	7.45
Romania	1,244	1,595	1,920	4.13	4.49	6.36
Eastern Europe	18,944	29,307	33,871	8.23	10.29	11.22

[a] Stock of FDI in June 1996.

Source: Austrian Ministry for the Economy, own calculations.

In early July 1995, the Slovak government surprised both foreign and domestic investors by abandoning plans to give shares in state enterprises to citizens in exchange for vouchers. In addition to this measure, privatisation through direct sale in the country does not always lead to capture by the highest bidder at open auctions. There are rumours that many such sales have gone to supporters of the prime minister Mr Meciar at give-away prices. Moreover, another regulation has excluded companies of 'strategic importance' from the privatisation process (Urban, et al., 1996). As a result of all these measures, net inflow of foreign direct investment into Slovakia stagnated at about 180 million USD in 1994 and 1995 (see Table 3). By comparison, the annual net inflow of foreign capital into the Czech Republic increased in the same period from 1,024 million USD to 2,720 million USD.

Mainly due to the disruption of the privatisation process and in contrast to the development in the Czech Republic, substantial foreign direct investment by potential investors has not taken place in Slovakia. This is the main reason why Austria's share of total foreign direct investment has remained relatively stable (or even increased slightly) since 1992. Table 4 shows the regional structure of foreign direct investment stocks in Slovakia as of December 1996. With a share of 21.0 %, Austria is Slovakia's largest investor, followed by Germany (19.8 %), the Czech Republic (16.4 %) and the USA (10.5 %).

The breakdown of foreign direct investment at the end of 1994 reveals manufacturing to have the largest share (49.3 %), followed by trade and services (25.5 %) and financial services (13.1). Investment in construction (1.6 %) and tourism (3.9 %) is shown to be of only minor importance (Slovak Statistical Office, 1996).

Table 5. Regional Structure of Foreign Direct Investment in Slovakia

	FDI stocks, December 1996	FDI stocks, March 1996	FDI units, Dec. 1995 Units	Share	share of stock/ share of units
Austria	21.0	21.4	1,380	15.8	1.4
Germany	19.8	17.8	1,088	12.4	1.4
Czech Rep.	16.4	16.3	1,614	18.5	0.9
USA	10.5	11.2	0,208	2.4	4.7
UK	NA	0.7	0,073	0.8	8.4
Netherlands	NA	6.2	0,145	1.7	3.7
France	NA	5.7	0,101	1.2	4.9
Others	NA	14.4	4,131	47.3	0.3
Total	100.0	100.0	8,740	100.0	1.0

Source: Slovak Statistical Office.

The breakdown of Austrian FDI in Slovakia shows above average shares in trade and services (+7.7 percentage points) as well as in financial services (+2.2 %). The share in manufacturing is well below the average (-6.5 %). When we compare the share of FDI stocks to that of FDI units (column 4, Table 4) we see that Austrian investors in Slovakia invest much less capital per unit than those from the US, the UK, or France. In addition, the high share of joint ventures among Austrian FDI indicates that these investments include a large number of small-scale, non-manufacturing investments. Austrian companies have some major advantages especially in the trade and services sector (financial and production-related services) due to geographical, historical and cultural proximity. So far, it appears these advantages have been put to rather favourable effect.

The regional distribution of foreign direct investment in Slovakia shows the absolute pre-eminence of Bratislava (see Table 6). Nearly 64 % of all foreign direct investment is funnelled into the capital. There was no other district with a share of more than 5 % in 1996. Districts with more than 3 % are distributed throughout the rest of the country.

Since only 8.3 % of the Slovak population live in Bratislava and the city embraces just 11.7 % of the country's workforce, the regional distribution of FDI is far from proportional. The massive concentration of FDI in Bratislava intensifies an already existing regional disparity in Slovakia, on the one hand between urban and rural areas, on the other between west and east. This disparity will show up in the following section as well.

Table 6. Regional distribution of foreign direct investment in Slovakia (in %)

Region	December 1996	December 1995	December 1994
Bratislava	63.7	62.1	60.9
Humenne	3.2	3.7	4.9
Nitra	3.3	2.7	3.4
Poprad	3.6	4.1	5.4
Prievidza	3.6	4.1	5.1
Ziar nad Hronom	3.8	4.4	0.4
Others	18.8	18.9	19.9
Total	100.0	100.0	100.0

Source: Slovak Statistical Office.

3.3 Labour Market and Labour Mobility

The labour market is the segment of the economy with the most direct bearing on the economic well-being of the population of a region or country. It determines not only the income the population has at its disposal but also impinges directly on the social welfare system through unemployment benefits, early retirement, etc. This also makes labour market issues vulnerable to political haggling.

Table 7. Unemployment Rates

Year	Austria	Vienna	Slovakia	Bratislava
1981	2.6	2.1	NV	NV
1983	4.5		NV	NV
1987	5.7	4.0	NV	NV
1990	5.4	5.8	1.6	0.9
1991	5.8	6.3	11.8	6.5
1992	5.9	6.4	10.4	3.8
1993	6.8	7.2	14.4	4.5
1994	6.5	7.1	14.8	3.9
1995	6.6	7.3	13.1	4.7
1996	7.0	7.8	12.8	4.1

Source: ÖSTAT, SSU.

Although the unemployment rate has been on the rise in both Austria and Slovakia (Table 7, Figure 4), the reasons for this have clearly been quite different. In Slovakia, unemployment has been rising as a result of a transition crisis marked by negative growth rates and high inflation. After four years of declining GDP, Slovakia is now experiencing high growth rates. According to a recent report, the growth rate of Slovak GDP was 6.8 % in 1995 (just below the 7.0 % reached in Poland) and 6.9 % in 1996 (the highest in Europe). As a consequence, the unemployment rate fell to 12.8 % in 1996.

With 7.0 % in 1996[5] the unemployment rate for Austria is only about half of that for Slovakia, but it has been increasing almost uninterruptedly since the 1970s, when the unemployment rate for Austria was regularly under 2 % (Richter, 1994). This increase in unemployment has been accompanied by growing rates of long-term unemployment and is understated to the extent that early retirement schemes have been used to keep people off the unemployment register. The latter have contributed to budgetary problems that led to across-the-board budgetary cut-backs in 1996.

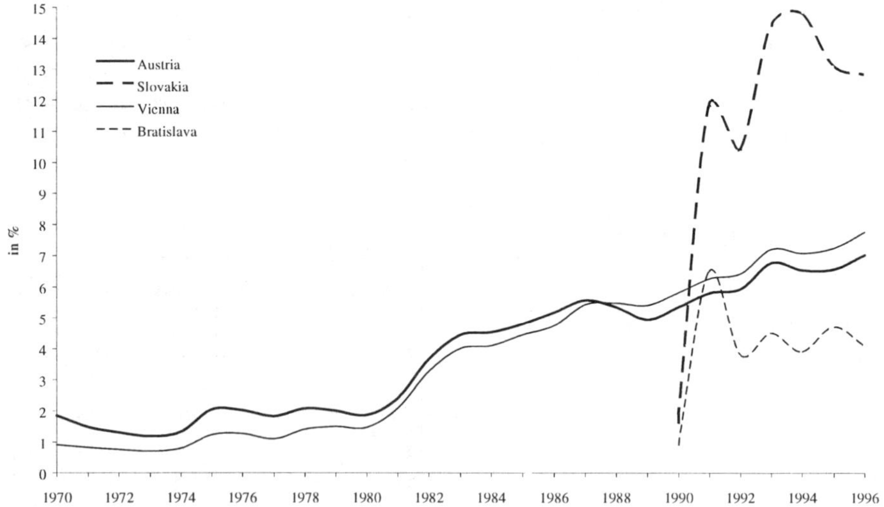

Source: IHS.

Figure 4. Unemployment Rates in Austria and Slovakia

[5] These figures are based on the number of people signing on for unemployment benefits. On the basis of EU specifications, the unemployment rate for Austria was just 4.1 % in 1996.

This situation is providing the backdrop for heated political debate on issues such as immigration, refugees and foreign labour. To a large extent it is a debate with a momentum of its own, and the views aired are often not supported by facts (Zuser, 1996). Despite a fairly positive general attitude towards refugees before 1990, positive public reactions to the changes in eastern Europe as they happened, and the economy's on-going need for foreign labour, sentiments to the effect that the labour market is being "flooded" have, together with requests for tougher restrictions, been becoming increasingly vociferous. Not only has this led to a significant shift in public opinion, there are now restrictive regulations in place for all (non-EU) foreigners in Austria, even those who have lived and worked in the country for many years.

As far as the regional distribution of unemployment is concerned, the situation for Vienna and Bratislava again differs. In the 1980s, unemployment rates in Vienna moved from below average to above average for the country. This was not a temporary phenomenon. In 1996 the unemployment rate for Vienna (7.8 %) was perceptibly higher than the Austrian average (7.0 %). This is usually explained by citing Vienna's peripheral location relative to the European market and a concomitant lack of competition that engenders less competitive industrial structures (Fassmann et al., 1991; Schremmer, Krajasits, 1992). Bratislava, on the other hand, is in a highly favourable position within Slovakia, up against the border with the west. Consequently, the city has posted the lowest unemployment rate of all Slovak districts throughout the transition period. In 1996, the average unemployment rate in Slovakia was 12.8 %, the figure for Bratislava just 4.1 %. This differs markedly to some districts in eastern and southern Slovakia where the unemployment rate is above 20 %. In fact, the unemployment rate is lower in Bratislava than in Vienna, turning the national figures on their heads.

There is still an enormous difference in per-capita-income between Austria and Slovakia. Monthly average wages (before taxation) in industry amount to only 8,583 SKK (about 270 USD) in Slovakia, while the average wage in Austria stood at 24,260 ATS (about 2,300 USD) in 1996. As in other transition economies, we can discern wages in Slovakia catching up only very slowly. Accordingly, the differences in nominal income can be expected to persist for some time to come. When taking into account purchasing power, the difference in GDP per capita is still high, though less dramatic (see Figure 5). According to Havlik (1996), Slovak GDP per capita in purchasing power parities had already reached 41 % of the EU15 average by 1995, bettered only by Slovenia (58 %) and the Czech Republic (52 %). Moreover, it expects to reach 63 % of the EU15 average by 2010. This would correspond to the income level of Greece. Austria's GDP per capita in purchasing power parities is expected to be stable at 13 % above the average of the EU15.

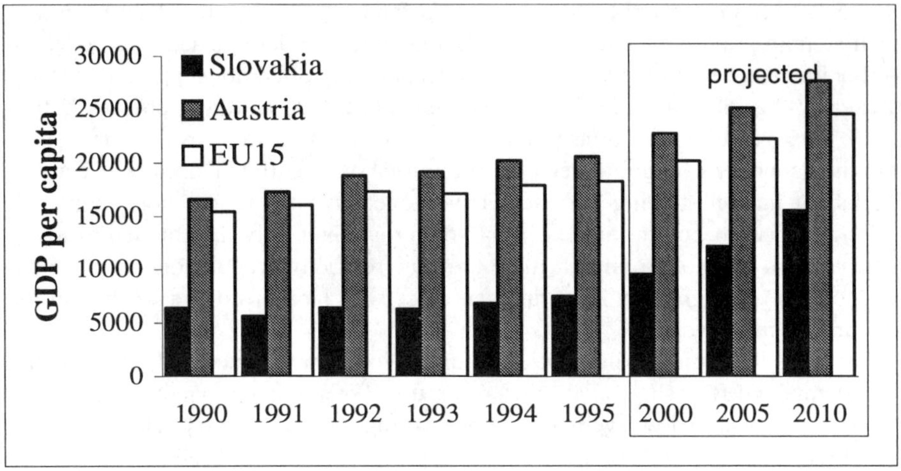

Source: Havlik (1996).

Figure 5. GDP per Capita in Purchasing Power Parities

The number of foreigners and foreign workers in Austria increased significantly in the late 1980s and early 1990s. As a proportion of the total Austrian population the figure rose from 3.9 % in 1988 to 6.6 % in 1991, the steepest increase of any OECD country (Biffl, 1992, p 526). The number of foreign workers doubled between 1988 and 1995, up from 150,000 to 300,000 (see Figure 6). The steepest increase occurred in the years 1989–1991.

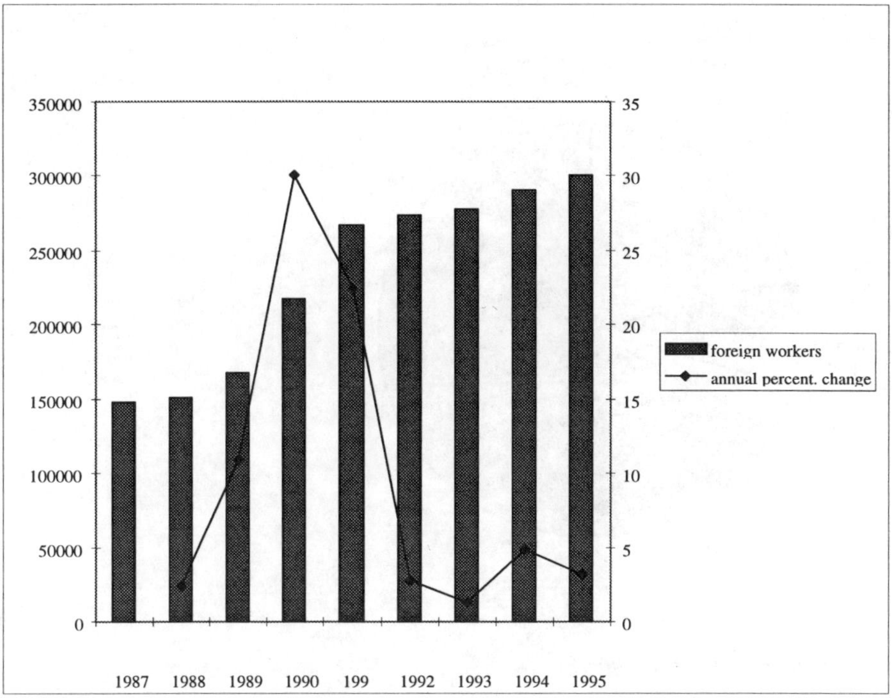

Source: Austrian Statistical Office.

Figure 6. Foreign Workers in Austria

This increase, however, cannot be attributed to the opening up of the Eastern Bloc countries. Sixty per cent of newly arrived foreign workers come from the traditional sources of migrant labour in Austria, viz. the countries of former Yugoslavia and Turkey. As Figure 7 shows, these still represent more than two thirds of the foreign workforce in the country. Labour migrants from the old Czechoslovakia[6] make up only about 3.5 % of all foreign labour, less than the figure for German or Polish workers.

It can be argued, of course, that these official data are incomplete since workers from Bratislava will most likely commute to Austria on a daily basis and may therefore be more likely to engage in short-term black-market employment. If this were the case, they would, however, need to use the established channels for finding such jobs and would need to cross the border between the two cities.

[6] Czech Republic and Slovakia are not treated separately in this statistic.

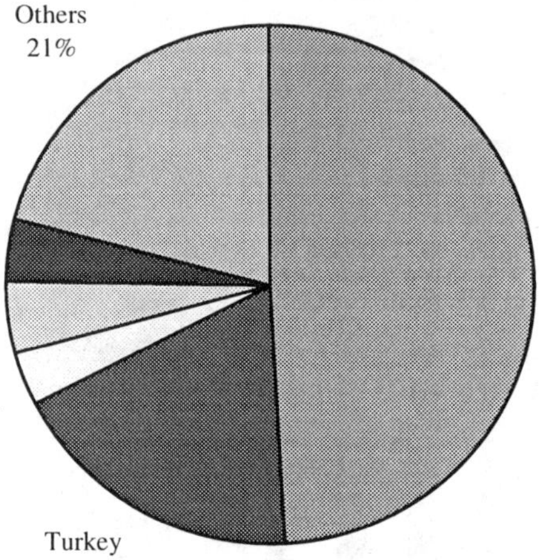

Source: Austrian Statistical Office.

Figure 7. Share of foreign workers in Austria 1995

Fassmann et al. (1993) analysed ads placed by foreigners looking for work in a local newspaper (Bazar) typically used for temporary recruitment purposes. They found that the share of ads from the old Czechoslovakia was high in 1990 and 1991 but declined considerably in the years thereafter. A follow-up analysis in 1994 showed that just 18.5 % of foreign job-hunters were Czechs or Slovaks. By contrast, the share of Poles was 48.3 %.

Maier (1994) reports the results of interviews conducted at the border stations between Vienna and Bratislava. He found that 39 % of all trips observed were interactions between the two cities. If the hinterland of the two cities is also considered, more than 62 % of interactions were accounted for. Of the 812 people going into Austria who were interviewed, only one in three stated that work was the purpose of their trip. Other purposes (business, shopping) were almost as important. Computations based on these survey data show that the number of Slovaks commuting to work in Austria can hardly exceed 2,500.

4 Summary and Conclusions

The three economic mechanisms that we have analysed in this paper – trade, direct investment, and labour mobility – are all operating as expected in the Vienna/Bratislava region. Trade has intensified tremendously in both directions, capital has flowed into Slovakia, and labour into Austria. The development is strongest with respect to trade, weakest with respect to labour mobility.

The level of trade between Austria and Slovakia has increased dramatically. Exports from Austria to the former CSSR/CSFR increased by 375 % between 1988 and 1996, imports by 239 %. The trade balance as viewed from an Austrian perspective has switched from a deficit of 1.7 billion ATS in 1989 to a surplus of 3.6 billion ATS in 1995. Such trade deficits, however, will not be sustainable in the long run for the CEE countries. In order to arrive at a more balanced exchange of goods, further trade liberalisation on the part of the EU and more competitive products on the part of the CEE countries will be needed. Import restrictions like those imposed by the Czech Republic can only be a temporary measure. The rise of intra-industry trade is one of the most striking features in the current pattern of trade evolving between the CEE countries and the EU. This indicates that a new division of labour is emerging surprisingly fast.

The analysis of foreign direct investments shows that this development is distributed rather unevenly. Foreign direct investments are concentrated in few countries of CEE and in these countries, as the example of Slovakia shows very clearly, in few regions. In Slovakia almost two thirds of FDI is concentrated in the city of Bratislava and thus at the western end of the country. In all likelihood, the stimulating effect of an inflow of foreign capital is being confined to the city and its immediate vicinity and is hardly being felt at all in the rest of the country.

This view is supported by our analysis of labour mobility and the labour market. The unemployment rate for Bratislava is way below that of the country as a whole and, indeed, is lower than that for Vienna. Labour mobility between Slovakia and Austria is surprisingly low despite substantial income differentials. To some extent this can be attributed to Austria's restrictive policy of attempting to purge the Austrian labour market of eastern European competition. However, we know from the theoretical analysis in section 2 that this attempt cannot be successful in the long run because of the side effects of trade and direct investment on the labour market.

Although the economic mechanisms that we have discussed in our paper seem to work in the direction one can expect from theoretical considerations, they have not lowered the development gap to any significant extent yet. There are still tremendous income disparities and this is unlikely to alter much for decades to come. This would appear to back up the polarisation theory of modern growth, which holds that development differentials will not be eliminated quickly.

Some of the mechanisms hampering the transfer of economic development into CEE countries may be outside the economic domain and hence outside the main focus of our analysis, though closely related to it. Reiterer et al. (1995), for example, analyse administrative cooperation between Vienna and Bratislava and identify numerous obstacles. Structural problems within the respective administration, lack of resources, political uncertainties, etc. all conspire to depress the number of cooperative ventures and joint activities as well as their success rate. To some extent, Reiterer et al. (1995) argue, it is the development gap that creates barriers and makes it difficult to collaborate on equal terms. Such mechanisms may actually help stabilise the development gap. In particular, the authors bemoan the fact that a cross-border development plan for the whole region is neither in place nor remotely in sight. It could well be that Vienna and Bratislava have already squandered the opportunity to establish themselves as a potent hub in the European system of cities and regions. Such deficits in cooperation and the persistent development gap indicate that one should still use the term region with care in the context of Vienna and Bratislava.

However, any more intensive integration strategy for the Vienna/Bratislava region may be fraught with major economic and political risks. We have shown that economic performance in Bratislava is startlingly better than that for the rest of Slovakia. Furthermore, we expect these disparities to increase in the period ahead as Slovakia is integrated into the EU. Despite significant positive spillovers into other regions, this development may further exacerbate political tension between Bratislava and less advantaged regions that is already significant. In this way, improved integration of the Vienna/Bratislava region may actually heighten internal instability in Slovakia, a factor which even now is often viewed as being set to jeopardise the economic development of central Europe as a whole.

References

Aiginger K (1995) Die Wirkung der Ostöffnung auf die Struktur der österreichischen Industrie, Bundeskanzleramt, Schriftenreihe Europa, pp 191–259

Altzinger W (1995) Beschäftigungseffekte des österreichischen Osthandels, 1988–94; Working Paper No. 34 of the Institute of Economics, University of Economics and Business Administration, Vienna

Barro RJ (1991) Economic Growth in a Cross Section of Countries In: The Quarterly Journal of Economics, vol 106, pp 407–443

Barro RJ, Sala-i-Martin X (1991) Convergence across States and Regions, Economic Growth Center, Yale University, Discussion Paper Nr. 629, New Haven/Conn.

Biffl G (1992) Auswirkungen des Ausländerzustroms auf den Arbeitsmarkt. In: WIFO-Monatsberichte, 10/92, pp 526–535

Borensztein E, De-Gregorio J, Lee, JW (1995) How Does Foreign Direct Investment Effect Growth?, NBER Working Paper, No. 5057

Breuss F, Schebeck F (1996) Opening up of Eastern Europe and Eastward Enlargement of the EU. In: Austrian Economic Quarterly, 2/1996, pp 97–107

Bröcker J (1994) Die Lehren der neuen Wachstumstheorie für die Raumentwicklung und die Regionalpolitik. In: Blien U, Herrmann H, Koller M (eds), Regionalentwicklung und regionale Arbeitsmarktpolitik, Konzepte zur Lösung regionaler Arbeitsmarktprobleme?, Beiträge zur Arbeitsmarkt- und Berufsforschung Nr. 184, Nürnberg

Collins S, Rodrik D (1991) Eastern Europe and the Soviet Union in the World Economy. Institute for International Economics, Washington, D.C.

EBRD (1994) Transition Report, Foreign Direct Investment, London

ECE (1995) Economic Survey of Europe in 1994–95, New York, Geneva

Fassmann H, Kohlbacher J, Reeger U (1993) "Suche Arbeit" – eine empirische Analyse über Stellensuchende aus dem Ausland, ISR-Forschungsberichte, Nr. 10, Institut für Stadt- und Regionalforschung, Österreichische Akademie der Wissenschaften

Fidrmuc J (1997) Effects of Integration of the CEEC into the EU on Austrian Trade Structure, Institute for Advanced Studies, Manuscript, Vienna

Fidrmuc J, Fidrmuc F (1995) Impediments to Slovakia, Institute for Advanced Studies, Vienna

Fidrmuc J et. al. (1994) The Slovak Republic After One Year of Independence. Study commissioned by the Bank Austria AG, Vienna

Grossman GM, Helpman E (1992) Innovation and Growth in the Global Economy. Cambridge/Mass.: MIT Press

Havlik P (1996) New Estimates of Aggregate Productivity and Unit Labour Costs. In: Podkaminer L. Slower Growth in Central and Eastern Europe, Delayed Stabilsation in Russia and Ukraine, WIIW Research Report No. 228, Institute for Comparative Economic Studies, Vienna

Hamilton CB, Winters LA (1992) Opening up International Trade with Eastern Europe. Economic Policy, vol 7

Heckscher EF, Ohlin B (1991) Heckscher-Ohlin Trade Theory. Cambridge/Mass.: MIT-Press

Hirschman AO (1958) The Strategy of Economic Development. New Haven/Conn.: Yale Univ. Press

Holzmann R, Zukowska-Gagelmann K (1996) "Exportchancen und Importschocks: Eine Untersuchung zur (A)symetrie der Handelsanpassung der OECD Länder im Gefolge der Marktöffnung in Mittel- und Osteuropa". In: Holzmann R, Neck R (ed): Ostöffnung: Wirtschatliche Folgen für Österreich, Manzsche Verlags- und Universitätsbuchhandlung, Wien

Krugman P (1991) Geography and Trade. Cambridge/Mass.: MIT Press

Layard R, Blanchard O, Dornbusch R, Krugman P (1992) East-West Migration - The Alternatives. Cambridge: MIT Press

Maier G (1994) Economic Integration in Central Europe: Cross-Border Trips in the Vienna-Bratislava Region, paper presented at the international seminar "Common Markets, Common Borders, Common Questions", June 6–7, 1994, Stadtschlaining, Austria

Maier G, Mašek M (1992) The Removal of the "Iron Curtain" and its Regional Impact upon Austria and the CSFR. In: Jahrbuch für Regionalforschung, Nr. 1991/92

Maier G, Tödtling F (1996) Regional und Stadtökonomik: Regionalentwicklung und Regionalpolitik. Vienna: Springer

Mundell RA (1957) International Trade and Factor Mobility. In: American Economic Review, vol 47, pp 321–335.

Myrdal G (1957) Economic Theory and Underdeveloped Regions. London: Duckworth

OECD (1995) The Regional Dimension of Unemployment in Transition Countries, Paris

OeNB (1995) Österreichische Direktinvestitionen im Ausland und ausländische Direktinvestitionen in Österreich, Statistische Monatshefte, 6/1995

Fassmann H, Findl P, Münz R (1991) Die Auswirkung der internationalen Wanderungen auf Österreich: Szenarien zur regionalen Bevölkerungsentwicklung 1991–2031

Reiterer A, Amesberger H, Halbmayr B (1995) Grenzüberschreitung Wien – Bratislava: Zur politischen Dynamik einer urbanen Zusammenarbeit. Vienna: Institut für Konfliktforschung.

Richter U (1994) Geographie der Arbeitslosigkeit in Österreich: Theoretische Grundlagen – Empirische Befunde. Vienna: Verlag der österreichischen Akademie der Wissenschaften

Romer PM (1986) Increasing Returns and Long Run Growth. In: Journal of Political Economy, vol 94, pp 1002–1037

Romer PM (1990) Endogenous Technological Change. In: Journal of Political Economy, vol 98, pp S71–S102

Schremmer C, Krajasits C (1992) Szenarien zur Ost-Grenzöffnung und deren Auswirkungen auf die österreichischen Ost-Grenzregionen. In: Schriften zur Regionalpolitik und Raumplanung, Nr. 20, Vienna: Bundeskanzleramt

Slovak Statistical Office (1996) several statistics

Stankovsy J (1996a) The Role of Foreign Direct Investment in Eastern Europe. In: Austrian Economic Quarterly, 2/1996, pp 109–120

Stankovsy J (1996b) Direktinvestitionen in Osteuropa: Österreich auf dem Rückzug? In: WIFO-Monatsberichte 5/1996, pp 349–354

UNCTAD (1995) World Investment Report 1995, Transnational Cooperations and Competitiveness, New York, Geneva

Urban W, Podkaminer L et al. (1996) Kräftiges Wachstum in Ost-Mitteleuropa, weiterhin Rezession in der GUS. In: WIFO-Monatsberichte 5/1996, p 355

Zuser (1996) Die Konstruktion der Ausländerfrage in Österreich: Eine Analyse des öffentlichen Diskurses 1990. Political Science Series No. 35, Vienna: Institute for Advanced Studies

Urban Networking as a Learning Process: an Exploratory Framework for Transborder Cooperation in the Baltic Sea Region

Perttu Vartiainen
Institute of Geography
University of Joensuu
Joensuu, Finland

The Baltic Sea Region: from a Transnational Vision to Local Action

The Baltic Sea Region is an area with a centuries-old tradition of trade and cooperation, the "Hanseatic League" being a particularly notable example of this (Joenniemi 1993). Its modern history is, however, full of contradictions and barriers to interaction both between the individual nation-states and, during the Cold War, between the western and eastern blocs. The process of political and economic transition underway in Europe since the late 1980s has dramatically changed this situation. The Baltic Sea Region is now an emerging European mesoregion whose integrative power appears to lie in new forms of (mainly maritime) transborder cooperation (Figure 1). The list of *potential* integrative forces is vast, stretching from economic relations and transport links to cultural relations and environmental management (Pedersen 1993, Joenniemi 1993).

Consequently, the essence of the Baltic Sea Region lies in the emerging scenario of a Europe of Regions and clearly constitutes more than just a growing cooperation among the states around the Baltic. The specific and most valuable results of Baltic cooperation will, in the long run, arise from its truly local and regional perspective (cf. Vartiainen 1997, Wæver 1993). It is the cities and population centres in the Baltic Sea Region that contain the most advanced and smoothly functioning physical and functional infrastructures. The functional core of the Baltic region is made up of the main cities around its coasts. In the three Nordic countries as well as in Estonia and Latvia, the main Baltic cities lie in the national core areas while in the other countries the 'Baltic dimension' constitutes, more or less, an antithesis to the national core areas and capitals (Figure 2).

Developments towards a higher degree of integration in the Baltic Sea Region are currently being intensified through a number of transnational initiatives. From an urban development point of view, the most important of these initiatives is "Vision and Strategies around the Baltic Sea 2010" (VASAB 2010), a cooperative framework for spatial development put in place by the ministries responsible for

Figure 1. Functional EURO-Regions (Veggeland 1994, 81)

planning in the Baltic Sea states (see also Graute 1996). In this initiative, 'cities and urban network' are seen as one of the three basic elements of spatial structure (together with 'mobility network and energy supply' and 'areas with specific problems and potentials').

Implementation of VASAB 2010 is now intermeshed with that of a number of schemes launched by the European Union, notably the new Interreg II C Community Initiative which is designed to address funding needs within the field of spatial planning, including transnational cooperation. Concrete measures in transborder cooperation are, in turn, supported through several Interreg II A programmes in the area as well as through the PHARE and TACIS programmes, which are designed to provide development support to the countries in transition from the former Eastern Bloc.

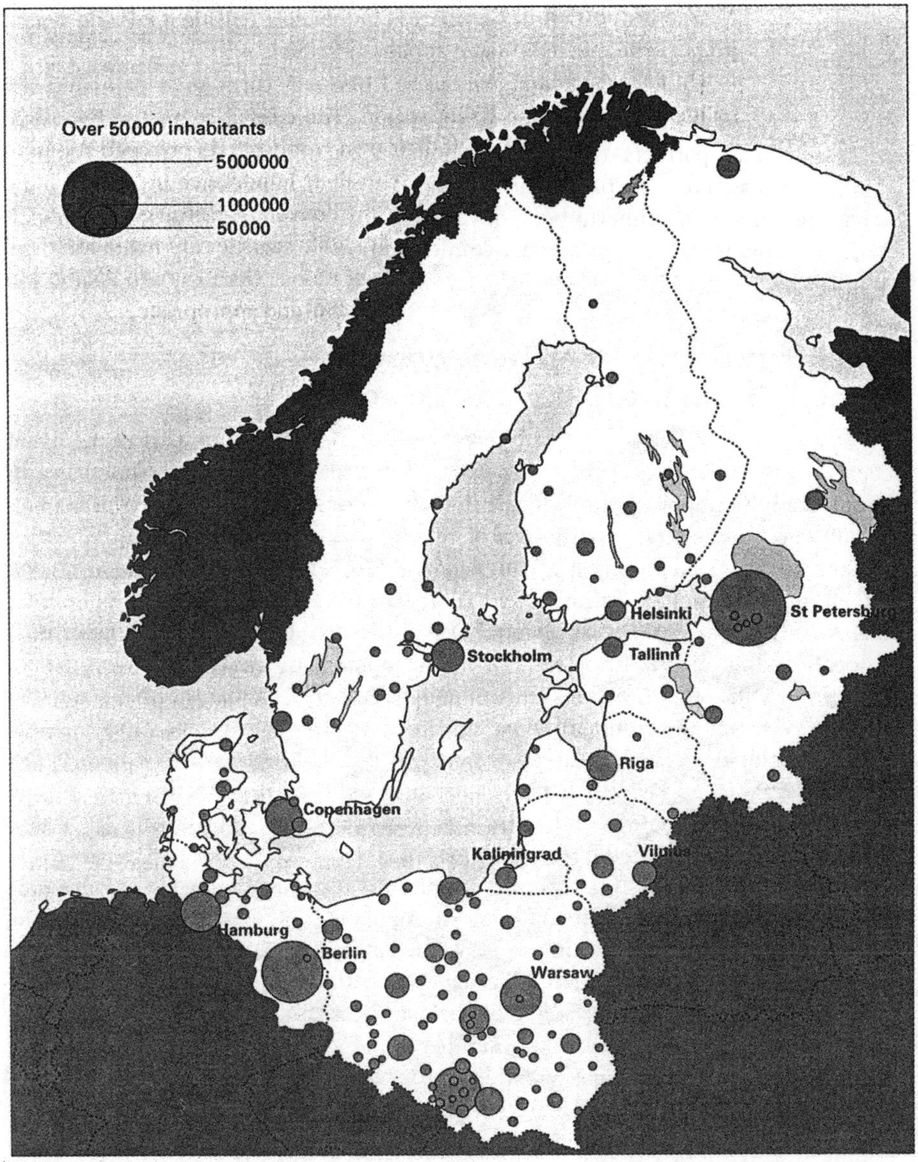

Figure 2. Urban Centres in the Baltic Sea Region (Kivikari 1996, map 2 based on VASAB 2010 excluding Belarus and Norway)

In this paper, I will develop the idea of urban networking in the Baltic Sea Region where cities are seen as active players in transborder cooperation and not simply as physical nodes in some greater urban system (Vartiainen 1994). The paper presents some basic ideas of the research programme 'Urban Networking as a

Learning Process in the Baltic Sea Region', which started life – based on a Finnish initiative (Forsström et al. 1996) – as a joint Nordic project in June 1996. Parallel to the 'communicative shift' in planning and the idea of learning-by-interacting, planning is understood as a learning process both locally and between the transnational partners (cf. Healey 1995a, Cooke and Morgan 1993). The learning perspective, in particular, emphasizes the process of learning-by-interacting, which has hitherto been developed as an extension of the networking paradigm in industrial studies and economic geography (see Morgan 1995, Dupuy and Gilly 1996).

Urban Networking as a Framework for Transborder Cooperation in the Baltic Sea Region

Recently several scholars in the field of spatial development in Europe have promulgated the importance of cities or urban regions as motors of economic development (Castells 1993, Bailly et al. 1996, for example). In this situation, cities and their regional coalitions are believed to compete more intensely with one another in an attempt to ensure their position in the spatial division of labour (Jensen-Butler et al. 1997, for example). What is meant by interurban competition are locally based efforts to promote the development of a city or city region in competition with other localities. Hallin and Malmberg (1996, pp 334–5) emphasize, however, that the competition of cities is used here mostly in a metaphorical way and regional wealth-creation is related to the degree to which firms in the region are able to upgrade their performance as opposed to attracting investment from outside. A competitive urban region is thus a region that is able to attract knowledge and apply it in the form of new economic activity.

One way of managing this complex situation is obviously *networking*, and that is why we are witnessing the mutual strengthening of urban networking and 'competition'. Urban networking is now an active practice in many European countries. In Germany, for example, a major experimental project has begun that is attempting to promote the concrete formation of urban networks in differing spatial configurations, e.g. around a number of large centres or between adjacent small or medium-sized towns and transborder areas (see Baumheier 1994, also Brake et al. 1996). The aim of these networks is to utilise and develop 'synergetic effects' by reinforcing interurban cooperation. This policy has been implemented through concrete measures in communications, infrastructure, tourism and culture, technology transfer, municipal management, etc.

We may differentiate between two basic points of departure when conceptualising 'urban networking':

1. As an economic and organisational principle, it implies mutually supportive actions or coalitions which are related neither through markets nor by administrative hierarchies. "A basic assumption of network relationships is that parties

are mutually dependent upon resources controlled by another, and that there are gains to be had by the pooling of resources" (Powell 1990, cited in Cooke, Morgan 1993, p 44). In the spatial development context, urban networking refers to local and regional cooperation between public and private organisations concerning business services, innovation, training, communications, etc.

2. In the basic physio-spatial sense, urban networking signifies polycentric urban configurations interconnected by linear infrastructures along which goods, people, information and money flow. Batten (1995, p 313) notes that "a network city evolves when two or more previously independent cities, potentially complementary in function, strive to cooperate and achieve significant scope economies aided by fast and reliable corridors of transport and communication infrastructure". This perspective places emphasis on strategic networks between logistic nodes (van den Berg, van Klink 1995).

In the same vein, Nijkamp (1993, p 157) distinguishes between physical and socio-economic networks. The first relates to infrastructure and telecommunications, the latter to interrelations occurring between a variety of players for a variety of purposes (for example, because of the interchange of knowledge, goods and/or information).

These two perspectives also appear to have surfaced, be it at a still more abstract level of spatial theory, in the recent Italian discussion on urban networking. The discussion has been summarised and taken further by Roberto P. Camagni for the most part. He believes that 'network logic' is replacing 'territorial logic', i.e. replacing the abstract Christallerian pattern of a settled hierarchy of centres and markets, as the main organising principle of inter-urban links. In this respect, the network operates as a 'club asset' providing benefits ('network externalities') solely to the members of the club, this being an intermediate structure between 'private' and 'public' goods (Camagni 1993, p 72). Camagni (1993, p 74) differentiates between two basic organisational forms of urban network: 1) complementarity networks made up of specialised and complementary centres and 2) synergy and innovation networks made up of similar, cooperating centres.

When studying urban networking from a policy perspective, an interesting extension might be a 'policy network' analysis (Heinelt, Smith 1996) that portrays networking as an organisational and economic principle as set out above (see also Cooke 1996, Dang-Nguyen et al. 1993). A policy network is seen to consist of different players – individuals, groups, organisations or institutions – interacting around specific policies or projects. The reason for maintaining networks is that network members have direct or indirect access to resources and influence in pursuing projects they consider to be individually or, more fundamentally, collectively, important. Policy networks are stable but permeable. There is some stability in these networks which derives from the establishment of trust and reliability, reputation and customary rules which members adhere to.

From the urban networking perspective, it is local government that is at the forefront of tackling the new spatial development challenges and social conflicts.

Reaching the practical player level, it must be emphasized, however, that urban networking should involve not only local government but also the other prominent public, private and intermediary players in the economic sphere, research and development, environmental management, etc. in their respective city or city region.

Theoretically, this leads us to discussions of different forms of urban governance and the capacity of leadership in 'global-local development' (see various articles in Healey 1995b, Judd and Parkinson 1990, for example).

Cities are also key players in transborder cooperation because it is only regions with an urban core that are likely to compete for new economic functions, an expanding fiscal base and greater administrative autonomy and they also form the context for the daily actions of both residents and economic institutions (Lever 1993, Robson 1992). In the Baltic Sea Region in particular, cities, being comparatively equal, are better vehicles for practical cooperation than states of very different sizes (cf. Wæver 1993, p 39). Nevertheless, the need for transborder cooperation often only relates to a clearly defined area or a group of localities.

In a transnational context, urban networking may refer either to cooperative arrangements in adjacent areas and between the nodes of a polycentric region or to non-adjacent alliances of cities with similar functions and problems. The transborder cooperation between local authorities in southern England and northern France – analysed by Church and Reid (1996) – is an example of the former while the pan-European networks and policy programmes of the European Union (Williams 1996, chapter 8) address the latter type. Chiefly from the latter perspective, Parkinson (1992, p 235) presents three functions of urban networks (specialising in one or more of these functions):

1. they act as a conduit for the transfer of knowledge and best practice, stimulating innovation in technology and policy;
2. they act as catalysts for the creation of economic cooperation between cities; and,
3. they lobby for national and, increasingly, European resources.

As Camagni sees it (1992, pp 372–3), transborder initiatives are more likely to have a positive effect if they involve regions with differing levels of economic prosperity rather than less prosperous border regions with similar problems. The Baltic Sea Region, and especially the actual cusp between East and West, clearly fulfils this condition.

Urban Networking as a Learning Process: some Exploratory Notions

In the research programme 'Urban Networking as a Learning Process in the Baltic Sea Region', urban networking will be analysed from the perspectives of the 'communicative shift' in planning and the learning-by-interacting concept. In the tradition of regional development and planning studies, these ideas have already been lent expression to in the writings of John Friedman (1987, for example). Next, I will briefly set out these basic ideas, and, then, some preliminary results of the first pilot of the above-mentioned research programme, the focus of which lies in the links between the Nordic and non-Nordic cities in the Baltic Sea Region.

Healey (1995a, p 52) has summarised the former idea under ten successive points. Planning emerges as:

1. an interactive and interpretative process;
2. undertaken *among* diverse and fluid "discourse communities", or cultures; and,
3. in ways, which require "respectful" interpersonal and intercultural discussion, searching out ways of recognising and valuing what we are trying to say to each other within the public realm.

This enterprise focuses our attention:

4. on the arenas where public discussion takes place and where problems and strategies are identified, evaluated and conflicts mediated;
5. on the multiple claims made for policy attention and the different forms in which these claims are made;
6. on ways of developing a critical, reflexive capacity which has both evaluative and creative potential, generating ideas about what to do and how, as well as testing them against the values of diverse community members; and,
7. on ways of opening out such strategic discourse in forms which are as inclusive as possible.

Through such new approaches to planning discourse, it is hoped not only

8. that participants will learn new things about themselves, their relations, their interests, values and understandings, but that, with such knowledge, they will
9. collaborate to change the way things are. In this way, such inclusive argumentation has the power to *transform* situations, through the power of better argument, and the power of ideas, metaphors, images and stories... Planning, as an explicit exercise in imagining the future, is thus about
10. "dreaming the possibility of change", imagining how to "start out on a journey" in mutually acceptable ways, rather than, as in the ideas of the urban designer planners, "dreaming the destination".

In Figure 3, I have presented this idea in more practical and formalised planning procedure form. The figure is an elucidation of the method of interactive planning developed by the Association of Finnish Local Authorities (Aronen 1996, p 14, cf. also Vartiainen 1996).

Description and analysis of previous strategies

Identification of key players and their strategies

Supply and demand of basic resources

Opportunities for joint action

Negotiations and/or seminars between the various players

Drafts of joint projects

Linking of units to the projects

Further definition of strategic projects

Decisions and implementation

Figure 3. The interactive planning process: a model approach for local development

This kind of interactive and interpretative approach may be viewed as a continuation of the collaboration initiated during the VASAB 2010 process, though now more at the interurban than the international level. In a transborder context in which the attempt is being made to bridge the former East-West divide, the learning-by-interacting approach is rather novel. The focus of this approach has been in localised learning by firms, but there are now a number of engaging studies in which peripheral regions too are being successfully integrated into global flows of information and knowledge – chiefly by public efforts – (cf. Huggins 1997).

Hitherto, most research on the Baltic Sea Region has drawn on studies of international relations or on descriptive statistics of the economy, population, interstate trade, etc. (Joenniemi 1993, Lundqvist and Persson 1995, Kukliński 1995). However, a case study of the Baltic Sea Region needs to address itself more incisively to the very different cultural, political and economic realities of the various parties. Most of the obstacles to be encountered in European transborder networking are represented in this region (cf. Table 1). The table constitutes a synthesis of various items of empirical research on transborder cooperation in Europe.

Table 1. Obstacles encountered in European transborder networking (van Geenhuizen et al. 1996 et al., table 5)

Transport	Natural obstacle Congestion at border crossings (overload/low service level) Lack of safety (criminality) Missing/deficient links in networks*
Socio-cultural	Language and vocabulary* Educational and income level Extent of ability to identify partners (mental map) Ideology
Economic-Institutional	Economic system (principles)* Monetary system (stability and convertibility of currency)* Legal and fiscal system* Property and ownership regimes* Trade borders and tariffs* Price level* High cost of network participation
Technology	Level of communication technology and logistics management* Professional level of personnel* Infrastructure and navigation standards (communication)* Standards and specifications in manufacturing
Political	Instability of political systems*

* Particularly important between western and eastern Europe

Special emphasis should be placed on the opportunities available to local government to pursue transborder cooperation. International initiatives, apart from the traditional twinning relationships, are a very new development even for the rather 'internationalised'[1] and proactive Nordic cities. In the transition countries, by contrast, local administration is still in its infancy. Over the past few years, there has been a growing interest in twinning arrangements in the Baltic Sea Region in particular (cf. Johansson and Stålvant 1996). The cities involved are still mostly in the phase of becoming acquainted with one another but some of these interactions may lead to more long-term involvement. The Baltic Sea Region also lacks the intermediary organisations so crucial to technology transfer and innovation networks (cf. Cooke 1996, Hassink et al. 1995 in the border region case).

The spatial reach of urban networking is still very restricted and reflects an effective partitioning of the Baltic Sea Region into different subregions (see also Euro-Verbund 1995, a proposal for potential action areas in the Baltic Sea Region). The Finnish cities, for example, are basically geared towards Estonia and St. Petersburg on the one hand and Germany on the other (it must be noted that in

[1] Referring here to the international nature of the economy, universities, and even the day-to-day life of the population, for example.

Germany the Baltic coastal area is only a minor policy item in comparison with the southern parts of the country).

Instead of truly interactive, horizontal cooperation, existing partnerships between local authorities seem to be still very much tied in with the hierarchical structures of state administration and characterised by one-way communication and resource transformation. This is particularly applicable to the relationships between the Nordic countries and the transition countries.

According to many eminent scholars in spatial development studies, such as Amin and Tomaney (1995), for example, it is not conclusively evident that a new economic-geographic order, with greater degrees of freedom for local or regional action, would be beneficial for the peripheral areas. Conversely, only proactive cities and regions have *any* chance in this world and ensuring their success requires action at a variety of spatial levels. Consequently, there is still a long way to go before the Baltic Sea Region could claim to be a 'third European core area', as mooted in a scenario of the potential Nordic impact on the evolving European spatial organisation (Veggeland et al. 1993). Even the biggest centres in the Baltic Sea Region are subordinate nodes within the global or, indeed, national hierarchy (cf. St. Petersburg; Eskelinen and Vartiainen 1996) or else they are still 'mere' national capitals in stature.

References

Amin A, Tomaney J (1995) The regional dilemma in a neo-liberal Europe. European Urban and Regional Studies 2:2, pp 171–188

Aronen K (1996) Development planning for communities. In: Making Future Now. Good practices in Finland. The Association of Finnish Local Authorities

Bailly A et al. (1996) Changing cities: restructuring, marginality and policies in urban Europe. European Urban and Regional Studies 3:2, pp 161–176

Batten DF (1995) Network cities: Creative urban agglomerations for the 21st century. Urban Studies 32:2, pp 313–327

Baumheier R (1994) Städtenetze - Raumordnungspolitische Ziele und Anforderungen an den weiteren Ausbau städtischer und regionaler Vernetzung. Raumforschung und Raumordnung 52:6, pp 383–391

van den Berg L, van Klink HA (1995) Regional competition, city marketing and strategic urban networks. In: Cheshire PC, Gordon IR (eds): Territorial Competition in an Integrating Europe. Avebury, Aldershot

Brake K et al. (1996) Städtenetze. Bundesforschungsanstalt für Landeskunde und Raumordnung, Materialen zur Raumentwicklung, Heft 76

Camagni RP (1992) Development scenarios and policy guidelines for the lagging regions in the 1990s. Regional Studies 26:4, pp 361–374

Camagni R (1993) From city hierarchy to city network. In: Lakshamanan TR, Nijkamp P (eds) Structure and Change in the Space Economy. Springer-Verlag, Berlin

Cappellin R, Batey PW (1993) Regional Networks, Border Regions and European Integration. Pion, London

Castells M (1993) European cities, the informational society, and the global economy. TESG 84:4, pp 247–257

Church A, Reid P (1996) Urban power, international networks and competition: the example of cross-border cooperation. Urban Studies 33:8, pp 1297–1318

Cooke P, Morgan K (1993) The network paradigm: new departures in corporate and regional development. Society and Space 11:5, pp 543–564

Cooke P (1996) Policy networks, innovation networks and regional policy: a review of the literature and an example from South Wales. In: Heinelt and Smith (1996)

Dang-Nguyen G et al. (1993) Networks in European policy-making: europeification of telecommunications policy. In: Andersen SS, Eliassen KA (eds): Making Policy in Europe. SAGE, London

Dupuy C, Gilly JP (1996) Collective learning and territorial dynamics: a new approach to the relations between industrial groups and territories. Environment and Planning A 28: pp 1603–1616.

Eskelinen H, Vartiainen P (1996) New rendezvous in St. Petersburg. In: Varis E, Porter S (eds). Karelia and St. Petersburg. From Lakeland Interior to European Metropolis. Joensuu University Press, Joensuu

Euro-Verbund (1995) Action areas in the formulation of the European Spatial Development Perspective. General priciples and case-study on the Baltic Region of Northern Germany. Study commissioned by the Federal Research Institute for Regional Geography and Regional Planning (Bonn)

Forsström B, Vartiainen P, Andersson H (1996) Urban networking as learning process in the Baltic Sea Region. In: Graute (1996)

Friedmann J (1987) Planning in the Public Domain: From Knowledge to Action. Princenton University Press, Princenton

van Geenhuizen M et al. (1996) Trans-border European networking: shifts in corporate strategy? European Planning Studies 4:6, pp 671–682

Graute U (ed) (1996) Vision and Strategies Around the Baltic Sea 2010 and their Relevance to Central and Eastern Europe. IÖR-SCHRIFTEN 17. Dresden

Hallin G, Malmberg A (1996) Attraction, competition and regional development in Europe. European Urban and Regional Studies 3:4, pp 323–337

Hassink R et al. (1995) Technology networking in border regions: case study of Euregion Maas-Rhine. European Planning Studies 3:1, pp 63–83

Healey P (1995a) The argumentative turn in planning theory and its implication for spatial strategy formation. In: Pakarinen T, Ylinen H (eds): Are Local Strategies Possible – Scrutinising Sustainability. Tampere University of Technology, Department of Architecture, Urban Planning, Publications 29

Healey P et al. (1995b) Managing Cities. John Wiley and Sons, Chichester

Heinelt H, Smith R (1996) Policy Networks and European Structural Funds. Avebury, Aldershot

Huggins R (1997) Competitiveness and the global region: the role of networking. In: Simmie J (ed): Innovation, Networks and Learning Regions? Jessica Kingsley, London

Jensen-Butler C et al. (1997) European Cities in Competition. Avebury, Aldershot

Joenniemi P (1993) Cooperation in the Baltic Sea Region. Taylor and Francis, London

Johansson T, Stålvant CE (1996) The twinning of cities. A code for neighbourhood cooperation in the Baltic Sea Area? A paper presented at Kieler Woche 23.–25.7.1996

Judd D, Parkinson M (eds) (1990) Leadership and Urban Regeneration. SAGE, Newbury Park

Kivikari U (1996) The Legacy of Hansa. Otava, Helsinki

Kukliński A (ed) (1995) Baltic Europe in the Perspective of Global Change. Oficyna Naukowa, Warszawa

Lever WF (1993) Competition within the European urban system. Urban Studies 30:6, pp 935–948.

Lundqvist L, Persson LO (eds) (1995) Baltic Regions and Baltic Links, nordREFO 1995:2

Morgan K (1995) The learning region: institutions, innovation and regional renewal. Papers in Planning Research No. 157, Department of City and Regional Planning, University of Wales, Cardiff

Nijkamp P (1993) Towards a network of regions: the United States of Europe. European Planning Studies 1:2, pp 149–168

Parkinson M (1992) City links. Town and Country Planning 61:9, pp 235–236

Pedersen J, Storm (1993) The Baltic Sea Region and the New Europe. In: Cappellin and Batey (1993)

Powell W (1990) Neither market nor hierarchy: network forms of organization. In: Straw B, Cummings L: Research in Organizational Behaviour, vol 12. JAI Press: Greenwich, CT

Robson B (1992) Competing and collaborating through urban networks. Town and Country Planning 61:9, pp 236–238

Vartiainen P (1994) Urban region as a geoeconomic and geopolitical unit in "A Europe of Regions". In: Braun GO (ed): Managing and Marketing of Urban Development and Urban Life. Dietmar Reimer Verlag, Berlin

Vartiainen P (1996) Alueellisen kehittämissuunnittelun uudet eurooppalaiset mallit ja vaiheittaisen neuvottelusuunnittelun idea. Kunnallistieteellinen Aikakauskirja 24:1, pp 85–92

Vartiainen P (1997) The advantages of a network of Baltic cities. A report for the Council of Europe, the Colloquy "The challenges facing European society with the approach of the year 2000: strategies for sustainable development of the northern States in Europe", Helsinki 22–23 May 1997

VASAB 2010 = Vision and Strategies Around the Baltic Sea 2010, 1994

Veggeland N et al. (1993) Impact of the Development of the Nordic Countries on Regional Development and Spatial Organisation in the Community. Final Report. Volume I Synthesis. Nordic Institute of Regional Policy Research, Copenhagen

Veggeland N (1994) Regionbygging over grenser i det nye Europa – aktører, arenaer og regimer, nordREFO 1994:6

Wæver O (1993) Culture and identity in the Baltic Sea Region. In: Joenniemi P (1993)

Williams RH (1996) European Union Spatial Policy and Planning. Paul Chapman Publishing, London

The Danube Waterway as a Key European Transport Resource

Peter Schneidewind
Austrian Institute for Regional Studies and Spatial Planning
Vienna, Austria

Introduction

To a very large extent, the river Danube lends the regions and states in central and south-eastern Europe their specific characteristics. Even though no commonly accepted notion of a Danube space exists as yet, the hydrographic catchment area of the river, which extends over more than 800,000 square kilometres, seems to have become the object of ever growing interest for a great number of disciplines, e.g. political, economical, ecological and cultural.

This reflects the diversity of the river's functions, of which freight transportation is by no means the least important. For centuries the Danube waterway has formed an important item of transport infrastructure, contributing significantly to the economic development of the regions alongside the river. For technological and political reasons, however, inland waterway transport on the Danube has historically been unable to keep pace with the dynamic development of rail and road transport. Even in the best years of navigation on the Danube the utilisation factor of the waterway lagged far behind its (theoretical) capacity.

Nevertheless, within less than a decade there have been several significant changes in the pattern of inland navigation on the river:

- The opening of the Rhine-Main-Danube canal in 1992 signified the forging of a direct link between the river Danube and the overall network of waterways in western Europe. Great expectations were and still are vested in this new transcontinental link, which connects western Europe to the Black Sea and the Danube region to the ports of Antwerpen, Rotterdam and Amsterdam (ARA).
- The socio-economic changes in the CEE countries after the breakdown of the COMECON system have had enormous repercussions for the transport sector, which had been accorded special status under central planning.
- Transport flows and transport volumes changed abruptly in the context of the transition process in these countries, with increased momentum expected in the wake of accession to the EU.
- The transport infrastructure is showing very severe shortcomings in almost all the CEE countries, making it necessary to assess investment priorities very prudently not only at national but also, and more pressingly, at European level.

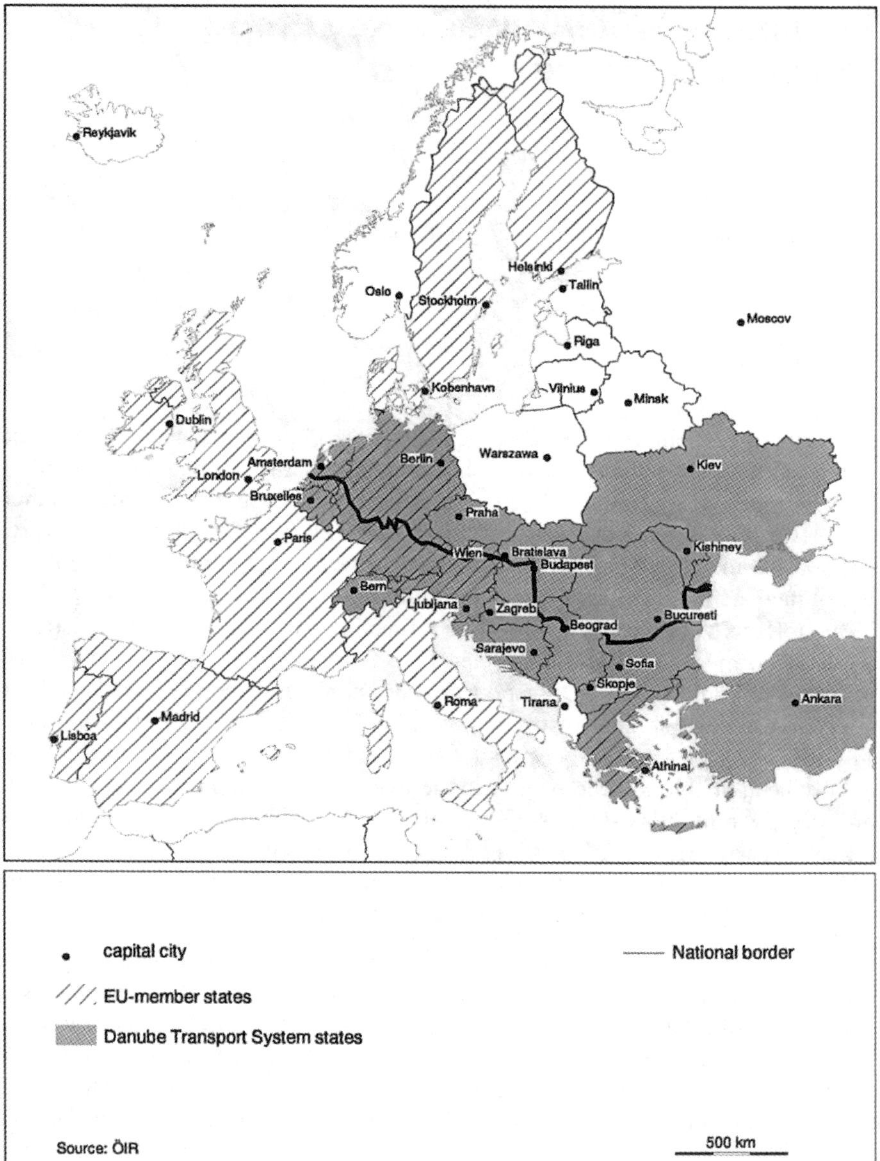

Figure 1. The states of the Danube waterway system

These considerations may have been in the mind of the European Commission when it decided to support a research project devoted to the Danube river transport system. The research proposal was put forward within the Fourth Framework

Programme for Research, Technology and Development (RTD) in the EU, which is the main source of European research funding

The proposing consortium is made up of just three parties (the Austrian Institute for Regional Studies and Spatial Planning [ÖIR] in Vienna, the European Development Centre for Inland Navigation [EBD] in Duisburg, and the firm Impetus Consultants from Athens) – three being the mandatory minimum number for Fourth Framework projects.

This partly reflects the fact that initially (i.e. in 1995) interest in the subject was limited. In the intervening period, however, the interest shown both by Western Europe (e.g. the Netherlands and Belgium) and, most notably, the 'accession' countries has significantly grown. And in the future phases of the project it should be possible to add participants from further Danube states. This development is made more likely still given that non-EU member states within CEE can now be integrated into EU research policy and hence gain access to the applicable funding.

The main objective of the research study is to provide a comprehensive evaluation of the technological, economical, and infrastructural aspects of the Danube waterway, hence the title of the study – and this article – from which the acronym EUDET (Evaluation of the Danube Waterway as a Key European Transport Resource) has been rather arbitrarily derived. This evaluation will eventually lead to the definition of conditions and measures for installing an efficient and attractive system of inland navigation for East-West European transport flows.

A second objective of the research project is to draw the attention of European decision-makers to the particularities of the Danube waterway. The Danube has hitherto played a rather minor role in European inland navigation, let alone in transport generally. It will be necessary to acquire appropriate information on the requirements and obstacles involved in switching significant volumes of cargo along the Danube corridor to conveyance by water to accord with the EU strategy of sustainable mobility of goods under the best possible environmental conditions as laid down in the Green Paper on the Impact of Transport on the Environment.

Finally the project sets its sights on creating and sustaining a network of stakeholders in the Danube waterway comprising operators and carriers, the transport industry, involved public authorities, research institutions and the media. This network covers in principle all states on which the Danube waterway system has a bearing, but is presently still most evolved in Central Europe.

The project started in January 1996. The research report will be finished in late 1997, most of the empirical information-gathering work having already been completed and documented in interim reports, one on infrastructure and fleets, the other on transport markets. A symposium entitled "Perspectives of Inland navigation in view of a wider Europe" organised by the EUDET consortium in Vienna this summer served as a first meeting place for a Danube transport research network in the making.

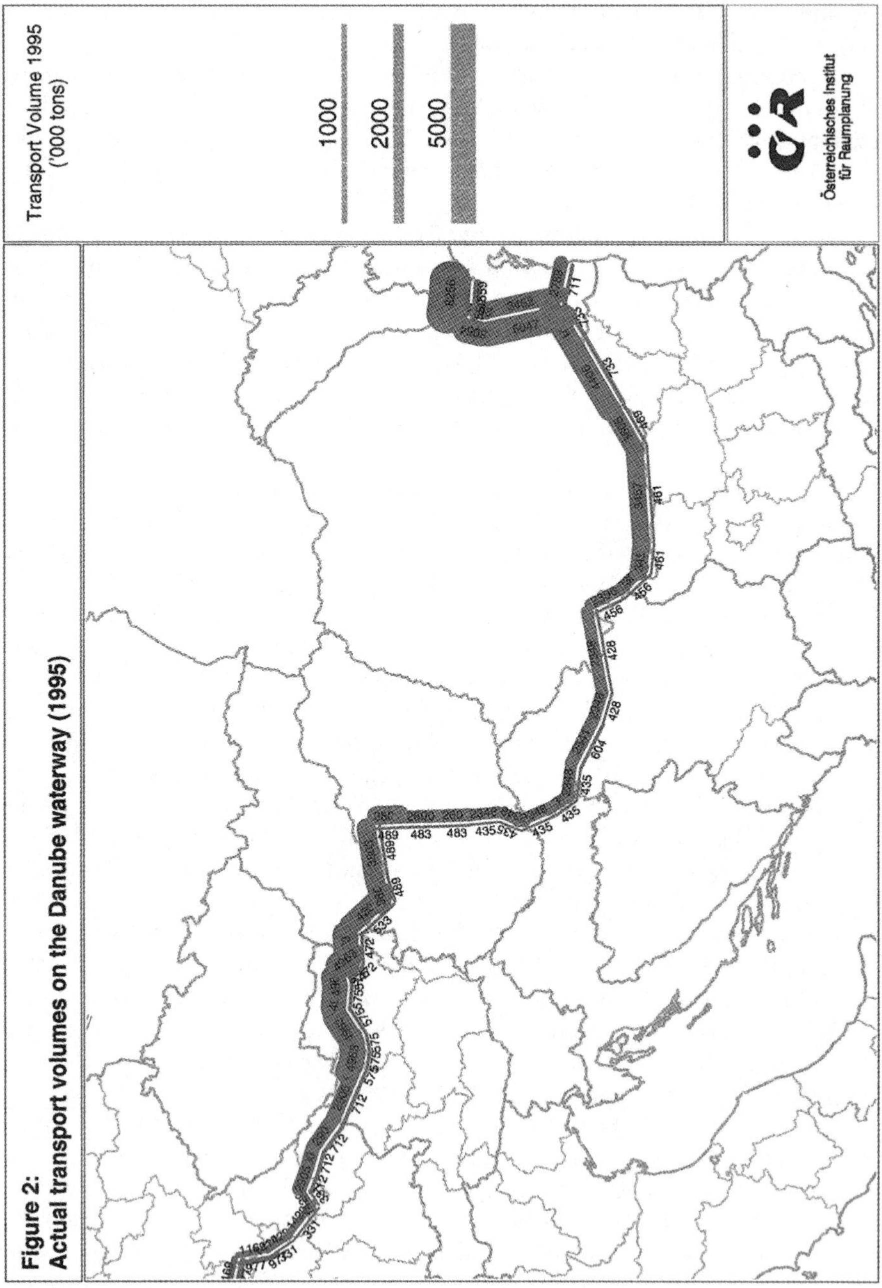

Figure 2. Actual transport volumes on the Danube waterway

Present Use of the Danube Waterway

One of the main research tasks involves giving an integrated view of the cargo volumes transported on the Danube and within the Danube Corridor and the Danube regions.

Drawing on different sources – mainly the Danube Commissions Port Statistics, national and EU transport statistics and foreign trade data – and using sophisticated software tools, one arrives at an overall picture of transport volumes as depicted in Fig. 2. The volume of transportation on the Danube in the riparian countries ranges from 1.9 million tonnes (Slovakia) to roughly 6 mt (Hungary 6.7 mt; Romania 6.5 mt; Austria 5.7 mt in 1995), adding up to 35 mt of cargo being transported on the Danube in one year (1995).

The foreign-trade component of this volume of transportation differs in structure in most of the countries concerned. On the upper Danube, cross-border traffic makes up the bulk of movements whereas in Hungary and in Romania internal traffic exceeds its external counterpart by far. This pattern is even more pronounced in Yugoslavia, where due to the UN embargo practically no international traffic at all was registered in 1995.

Except for Romania, the commodities transported internally in the riparian countries are almost exclusively gravel and sands dredged from the river bed.

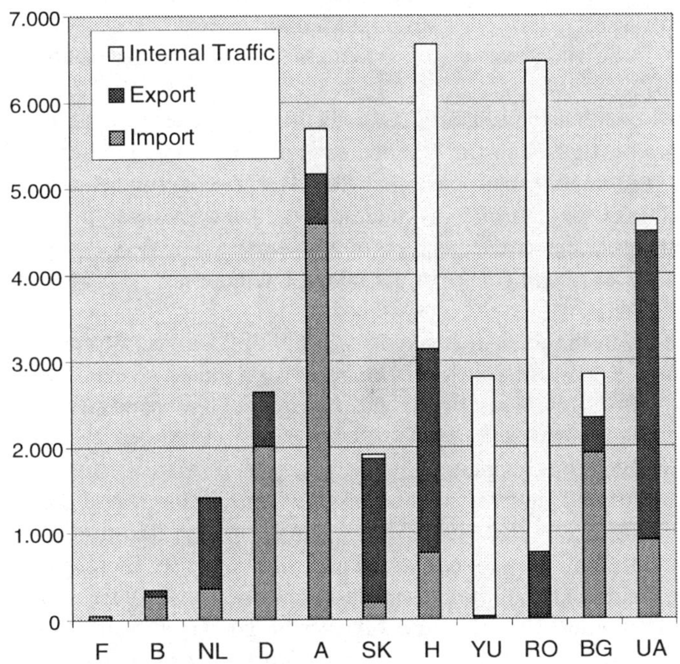

Figure 3. Transport volumes by country

The importance of long-range, cross-border traffic on the Danube is best shown in Fig. 4, which gives the results of the aggregation of relational transport flows. It can be seen that the biggest transport volumes are conveyed from the Central Danube region (Slovakia and Hungary) to the Upper Danube (Austria and Germany) and from the Black Sea region (Moldova, Ukraine, Russia and Turkey) to the Lower and Upper Danube.

These figures are the upshot of various developments over the past five years. In Austria transport volumes had started to grow even before the lifting of the embargo on Yugoslavia and in 1996 had returned to pre-1989 levels. German Danube transport volumes have also been slowly increasing, whereas in most of the transition countries the amount of goods shipped along the Danube has remained at the level to which it plummeted after the COMECON system broke down. As well as causing the bottom to fall out of the market in this region, the war in Yugoslavia also led to the UN embargo, which from 1993 to 1995 put a ban on goods being transited through Yugoslavia.

The first resolution restricting transport operations was adopted in May 1992 and prohibited transports to and from Yugoslavia. In April 1993 this was reaffirmed with an added stricture prohibiting the transit of goods along the Danube.

These resolutions were complied with, and this together with the ravages of war hindered all conveyance along the Danube. As a result, traffic from the Danube and other forms of carriage was switched to alternative routes through Romania and Bulgaria or else via Italy, crossing the Mediterranean.

In November 1995 the sanctions were suspended and in October 1996 they were lifted altogether.

It must be pointed out that the embargo was not the main cause of the downturn in inland navigation figures in the Danube region. The dramatic changes in the relationship between supplier and customer that overwhelmed industry in the region when the COMECON regime disintegrated predated the war in former Yugoslavia. Nevertheless, the consequences of the embargo may well be adversely affecting efforts to regain shares in the inland navigation market in – all – Danube countries.

At the same time, even the present transport volume of 35 million tonnes represents a modal split (i.e. inland navigation's share of all goods conveyed) within the relevant corridor of 7.4 %. This figure was highest for the section between Bratislava and Linz, where it stood at approx. 20 %.

Even more remarkable would appear to be the fact that, in the traditional bulk cargo segment, two thirds of the total volume of shipments in the Danube Corridor went by water. This means that within this segment, which traditionally accounts for more than 90 % of inland navigation cargo, the scope for increasing the market share is very limited and additional transport volume can only be expected from rising demand for raw materials and other bulk goods or from gains in commodity sectors that have so far not shown much affinity for Danube navigation.

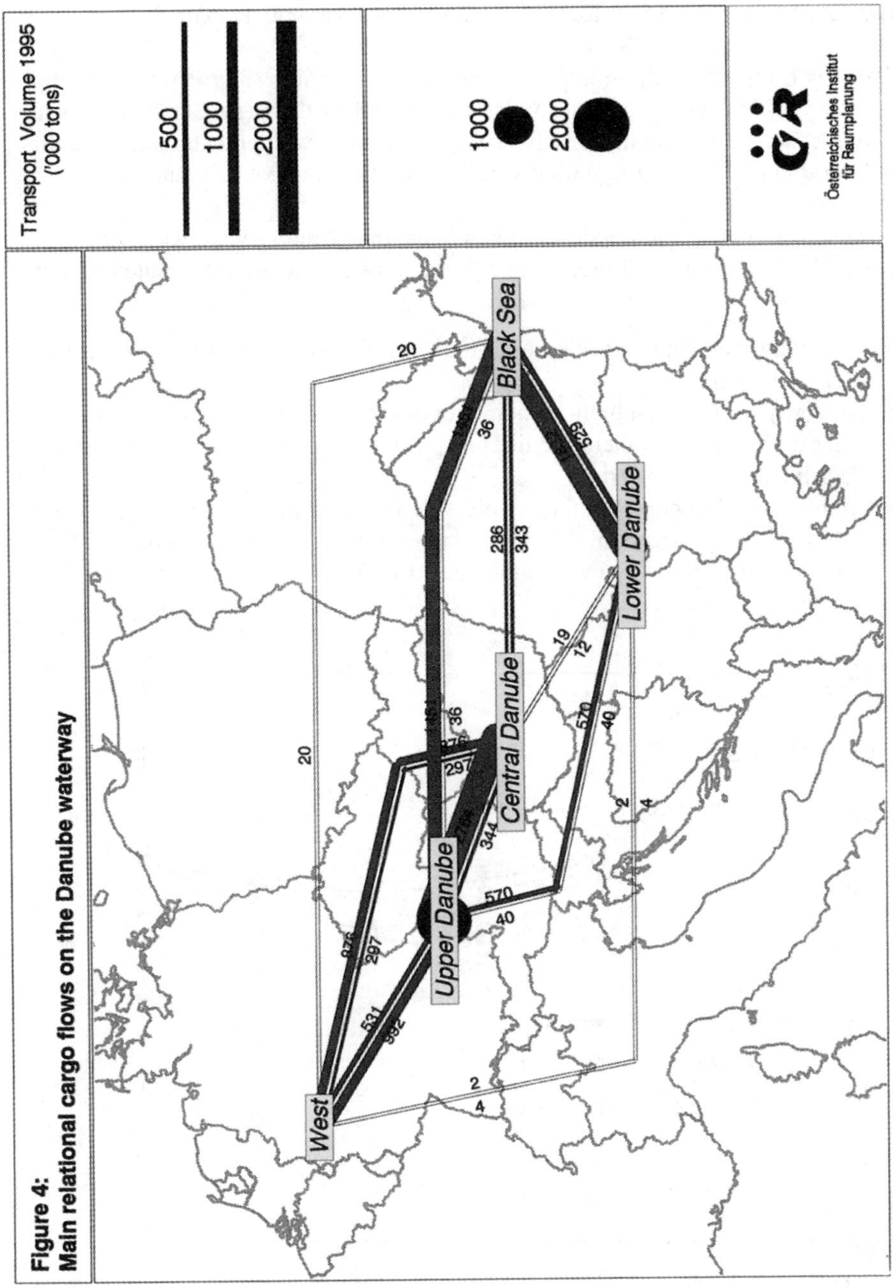

Figure 4. Main relational Cargo Flow on the Danube waterway (1995)

Considerations on the Future of Danube Transportation

There is no question that the present use of the Danube waterway leaves a lot to be desired both with regard to transport demand in the region and to the waterway's inherent potential. Clearly, though, future developments can be defined neither quantitatively nor qualitatively by merely citing past circumstances on the Danube.

When assessing future developments on the Danube waterway, one has to consider at least three different aspects, which of course are interrelated to a great extent:

- Firstly the development of foreign trade within the region and between the region and outside.
- Secondly the establishment of homogenous water-depth conditions and reliable and efficient transhipment facilities in ports, i.e. infrastructure investment, is of decisive importance and
- Thirdly developments on the supply side of the inland navigation market will determine how the possibilities of increased trade volumes and more adequate infrastructure are put to effect in the case of Danube-based navigation.

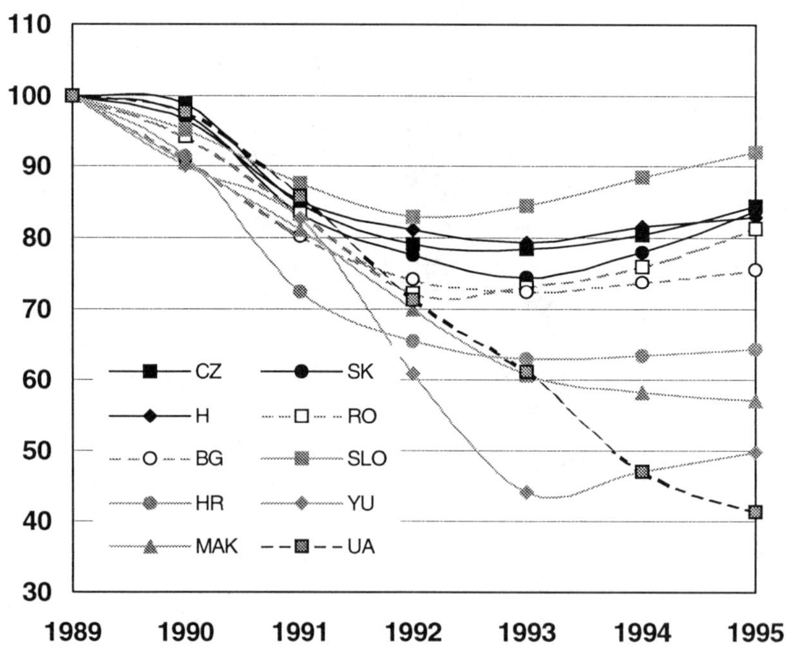

Figure 5. Growth in GDP in selected transition countries
Source: WIIW (1995, 1996, 1997).

Foreign trade has undergone changes of the most dramatic nature in the region over the past 5–10 years. The breakdown of the COMECON trade system has been accompanied by a comparatively fast reorientation of trade flows in the CEE countries from East to West.

For most of the countries in transition, economic growth has only begun to pick up recently. None of the Danube countries in CEE had reached its pre-1990 output level by 1995 (see Fig. 5). Trade volumes – both in money terms and by weight – did fare better. The pattern of intraregional trade changes in the period 1990–1995 is shown in Fig. 6.

Doubtless this pattern will not be sustained since it is dominated by the early phases of transition and the Yugoslavian war. But an analysis of these data reveals two factors whose significance for inland navigation is set to persist if not actually grow:

a) Increasing unit values in foreign trade on all routes – notably to and from overseas destinations. This reflects the growing importance of high-value commodities in trade but at the same time the decreasing relative weight of goods traded, which calls one of the main advantages of inland navigation into question. In the western countries trade growth nominally amounted to some 5 % p.a. in the period 1990–1995. But, significantly, this growth in value did not result in an increase in overall tonnage.
b) In all countries analysed the share of foreign trade accounted for by raw materials has been dropping to varying degrees with regard to both import and export operations.

It is thus overly simplistic to expect growing transport volumes on the Danube to materialise – as is claimed – merely from a growth in income and economic performance and induced foreign trade expansion in the Danube countries. The growth potential of the prime commodity sectors in inland navigation at present – mainly ores, minerals, chemicals and other bulk raw materials – is very limited even in the medium to long term.

Even more than on other waterways, inland navigation on the Danube can probably only be successful, i.e. maintain or increase its share of goods transported, if it turns to higher-value commodities. As can be seen from first experiences on the Upper Danube there is ample room for such reorientation, and the Lower Danube with its sea ports is likewise going to follow suit sooner or later as regards containerised transports and roll-on/roll-off services.

With some justification the adverse navigating conditions of the Danube waterway are considered to be the greatest obstacle to its more intensified and efficient utilisation. The water-depth in particular – the hydrologically most influential factor for passage through confined waterways – creates considerable drawbacks to navigation.

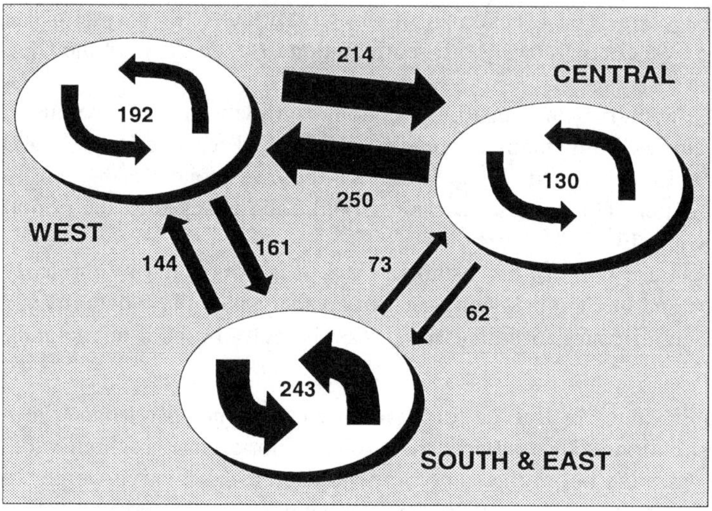

Figure 6. Change in volume of trade – a) by value (1990 = 100)
Source: UN World Trade Data Bank, own calculations.

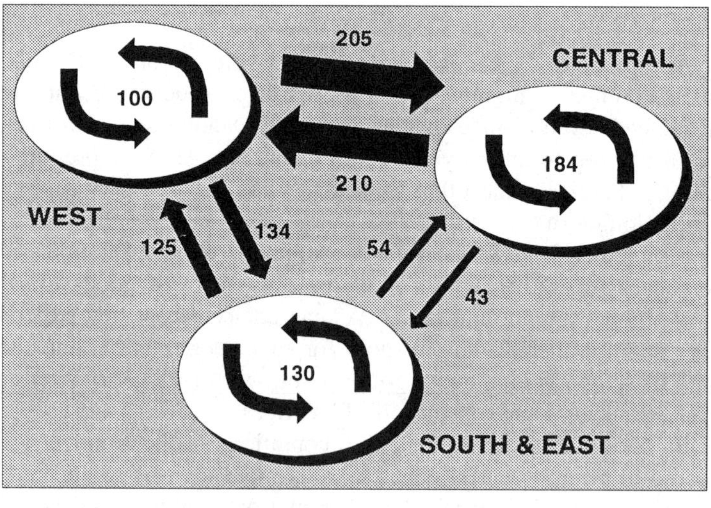

Figure 6. Change in volume of trade – b) by weight (1990 = 100)
Source: UN World Trade Data Bank, own calculations.

There are four sections of the river with frequently unfavourable depths of water:

The 69-km stretch between Straubing and Vilshofen (Danube km 2,318–2,249) can be treated as the most critical. On certain spots of this section upstream of Deggendorf (km 2,285) there is only an 89 % probability that the water-depth will even reach 1.70 m, and between Deggendorf and Km 2,247 of the Danube a depth of 2.00 m is ensured with the same probability (on 326 days of the year). The depth of water at LNRL[1] (Low navigation and regulation level) originally recommended by the Danube Commission, to wit 2.70 m and 2.80 m for bedrock soil, cannot be attained for 265 days a year. The Danube Commission accordingly scaled down its recommendation for the pre-canalisation stage to 1.85 m and 1.95 m respectively.

These water-depth conditions severely shackle the reliability (need to lighter, break of navigation) and efficiency (significantly lower draughts mean reduced profitability, market appeal, and flexibility in operation) of westbound Danube navigation as regards both conventional cargo and combined traffic. Unsatisfactory depths of water also impair the interoperability of the Danube and Western European waterways through the Main-Danube canal.

It has not yet been possible to implement existing plans to overcome this problem due to long-running discussions between experts on alternatives and strong environmental opposition.

On the Austrian section of the Danube waterway the navigable bottlenecks owing to insufficient water depths are the "Wachau" sector (km 2,038 and 2,008) and downstream of Vienna with shallow stretches down to Bratislava in the Slovak Republic (km 1,920–1,875). Apart from these two critical sections, the Danube waterway in Austria provides a minimum water-depth of 2.80 m, largely as a result of a chain of hydro-power plants and their reservoirs. This accords with the recommendations of the Danube Commission.

---

1 The Danube Commission (DC) originally released recommendations in 1960, and for the Upper Danube in 1962, (Document CD/SES 20/27) concerning characteristic dimensions for the waterway and for other hydraulic structures on the Danube ("Recommendations relatives á l'établissement des gabarits du chenal, des ouvrages hydrotechniques et autres sur le Danube", Budapest 1988). Member states thereby agreed that the improvement of the waterway is to be carried out in two steps and they undertook to assure minimum depths of water (Profondeur minimal /Mindestfahrwassertiefen) at LNRL for the phase preceding canalisation of the river (phase 1) as well as for the period following canalisation (phase 2). For the section Regensburg–Vienna minimum water depths have been fixed at 2.0 m and 2.1 m in the case of bedrock sole for phase 1 and at 2.7 m/2.8 m for phase 2. In one of its following sessions, the DC singled out the German section Regensburg–Kachlet after having found out that conventional hydraulic engineering measures did not produce the intended waterway depths and reduced the target parameters for the pre-canalisation phase to 1.85 m/1.95 m. Recommended water-depths for the second phase remained unchanged.

In the Wachau river bed, engineering measures carried out in 1986/87 have improved water-depths by restoring them to 2.50 m at LNRL. It is expected that regular maintenance work can prevent a worsening of the situation.

According to the recommendations of the Danube Commission, the navigable water-depth downstream of Vienna should be 2.50 m (in the first phase without canalisation). As recently conducted investigations into the navigable conditions of this sector have proved[2], the average depth of water available over the past decade was just 2.25 m. Thus, according to the hydrographic data, over the past two decades the section of the Danube east of Vienna has been unnavigable at 20 dm draught on an average of 84 days and at 25 dm draught on an average of 155 days, which reduces carrying capacity significantly.

It has to be stated that, whilst other European river systems have either officially guaranteed minimum depths of water or defined depths of immersion (sum of draught and sinking depths), or both, the Danube has neither. Hitherto, the draught of a vessel has been determined by the navigation company on the basis of water level data supplied by hydrographic service units. However, this does not mean that the draughts navigated are adjusted to actual depths of water in accordance with hydrographic data, but that these correspond to a standard depth of immersion valid for the lowest daily water level for the relevant month.

For example, until the dam at Freudenau was built the standard depth of immersion for the Vienna–Bratislava route was + 8 dm gauge reading at Vienna/Reichsbrücke. At LNRN, the draught measures 9.5 dm + 8 dm = 17.5 dm. Since a free water depth (safety distance to river bottom) of 5 dm is assumed, this gives rise to a navigable water depth of 22.5 dm for the minimum passage between Vienna and Bratislava at LNRN.

Furthermore, advance estimates of allowable draughts have to include a safety margin and must also take into account the time needed for the ship to reach the shallow spot ("forecast of permanent water level lines"). Applying this dynamic approach, the number of days with restricted draughts for traffic against the current (which cannot float along with the water conditions) is thus much higher than if calculated according to the permanent water level lines of waterway depths.

In its *Binnenschiffahrtsmemorandum 1992*, the Austrian Government renewed its obligation to assure fairway conditions east of Vienna which will allow a draught of 2.70 m. Intensive research into methods conducive to achieving this goal have been carried out in recent years, notably into conventional low water regulation methods combined with the continual addition of gravel and water regulation methods combined with artificial paving of the sole.

Very much as with the situation in Bavaria, no sure solution as to how to improve this particular section has yet been found.

---

[2] Teilgutachten Schiffahrt, Nationalparkvorbereitungskommission Nationalpark Donau-Auen, ÖIR, Vienna 1995.

Further downstream, traditional shallow spots at low water levels have been eliminated through the presence of the Gabcikovo complex and its navigation channel, which ensures water-depths of 3.50 m at LNRL.

From Palkovicovo (km 1,811) down to the Ipoly mouth (km 1,708) navigation is hampered at several points. On the section between Palkovicovo and Gönyú there are 6 to 7 bottlenecks where the navigation channel is shallow and narrow, owing to (sand) bars. At other locations the bed of the navigation channel consists of gravel and/or márga and the channel is restricted in depth.

Within the Hungarian section from the Ipoly mouth (km 1,708) to Budapest (km 1,635) the most unfavourable conditions for navigation occur at Nagymaros (km 1,697). The Nagymaros threshold may be considered as a critical shallow point, yielding depths of just 1.4–1.5 m for some periods of the year. A draught of 2.5 m cannot be guaranteed for 160 days of the year. These conditions have worsened gradually over recent years in the zone of the circular dam.[3]

In 1992, the Hungarian government proposed a three-step programme to improve navigation between Estergom and Vác which will not even come close to meeting the requirements of the Danube Commission regarding draught and availability before its third phase, which in turn is not scheduled to commence until after 2005. The first phase, originally scheduled to run from 1993–1996, stipulates a minimum depth of water of just 2.10–2.30 m for a navigable channel 80 m wide on an average of 240 days a year. Phase 2 is to deliver 2.50 m within the same parameters by 2005.

To conclude, the Danube upstream of Budapest does not satisfy the requirements of punctual, reliable and efficient inland navigation. Capital expenditure on improvements will not be effected before the year 2000, and even then it is not clear which method will be applied and by extension how much it will cost. Until such time, forwarders cannot be offered the full efficiency of inland navigation and hence will stick or turn to other modes – primarily road.

Whereas upstream of Budapest it is the waterway itself that needs improvement, downstream it is rather the port infrastructure that is in urgent need of modernisation (plus a need to actually build new ports). Inland navigation will have to become far better integrated into intermodal transport chains if it is going to fulfil its function as an energy-effective long-range transport mode for more than just bulk cargo.

This is not only true of port facilities in the narrower sense (hardware) but applies especially to the communications sector, in which Danube-based navigation seems to be lagging behind the other European transportation modes to an appalling extent.

[3] Programme of Development for the Infrastructure of Inland Navigation, Bureau for Research Organisation and Development at the Institute for Transport Sciences, Ltd., Budapest 1994, for the Ministry of Transport, Communication and Water Management.

This leads to the last of the three groups of considerations, the development of the competitiveness of inland navigation in general and on the Danube in particular. Despite the state of the infrastructure (bottlenecks etc.) and actual transport demand levels (commodity-specific), there are still a wide range of future development options. Besides general trends in national and European transport policy, i.e. restriction and regulation of various modes, it is the performance of competing and cooperating transport modes and transport service enterprises that determines the actual volume of transportation for a particular carrier/way. The relative position of Danube-based navigation in most of the Danubian countries (with Germany probably being the only exception) is not a favourable one.

Ship operators are mostly traditional state-type enterprises with little incentive to innovate and in the present climate virtually no financial resources of their own to invest. The situation is somewhat better with regard to the ports, most of which go far beyond their waterway-related activities by making industrial development premises available and engaging in non-navigation transport services. This is also beginning to be the case in the CEE countries.

It is important to note, that the relative cost position of inland navigation is much more dependent on the efficiency of the transhipment point, e.g. ports, than is generally imagined. Even for long distances the cost of transhipment can still represent a sizeable fraction of the overall cost and hence be an important factor in making a rational choice.

This is of particular importance for two reasons:

a) There is much more scope for fluctuation in transhipment costs than in waterway costs. This applies not only to monetary values but even more so if one considers time, reliability and safety as generalised cost factors.
b) The relative importance of the cost of transhipment increases in line with the unit values of the commodity transported. As has been pointed out, this increase in unit values was and will be one of the most significant developments in goods transportation.

Within the EUDET-project, a simulation model has been developed for forecasting transportation flows on the Danube, using simple cost functions for different modes. Leaving the present demand structure in the Danube space unchanged. This model shows much higher elasticity of Danube waterway transport volumes to transhipment costs than to waterway costs. For example, a decrease of 40 % in waterway costs would cause an increase of 27 % in transport volumes whereas the same reduction in transhipment costs would produce an increase almost twice as big (46 %).

These rather theoretical simulations of goods transport volumes will be further detailed and tested within the following months. In real life, though, competition does not take place between cost functions but between enterprises. This makes it often difficult to find the right balance between competition and cooperation. In the case of inland navigation strategic cooperation between key operators and a wide range of ports – both of different nationalities – and sometimes even with the competing modes with their feeder and/or distributive functions is of decisive importance. In most countries the ports will have to take a lead in this strategic cooperation – possibly fostered by public authorities.

Danube Transportation as a Testing Field for Transnational European Cooperation

Which way Danube-based navigation is going to develop will depend very much on the intensity and dimensions of cooperation among the various players within the Danube transportation system, which includes government, national transport industries and other non-governmental bodies. So far, cooperation among the states in question has not delivered much in the way of tangible results in response to the new situation for Danube navigation brought about by the political and economic changes in CEE and by the splicing of the two waterway systems Rhine and Danube via the Main-Danube Canal. On the other hand, the time span of little more than five years does not seem to be a sufficiently long time for the achievement of such results.

Historically speaking, Danube-based navigation can claim a very good and long-running record of cooperation, promoted mainly by the Danube Commission.

The Danube Commission was established at the Belgrade Convention signed in August 1948 by the governments of Bulgaria, Romania, Ukraine, Hungary and what were at that time Yugoslavia, the USSR, and Czechoslovakia. Currently, the participating states of the Danube Commission with its seat in Budapest are Austria, the Slovak Republic, Hungary, Yugoslavia, Romania, Bulgaria, Ukraine and Russia – all with full membership status. Germany, Croatia and Moldova have observer status. The Danube Commission can merely table recommendations relating to the technical and legal matters on the Danube waterway, traffic regulations, naval equipment etc. These recommendations and resolutions are not binding on the national administrations of member states, who are entitled to set different rules for services and traffic policy in the sectors under their own jurisdiction.

In recent years a number of new intergovernmental bodies dealing with the Danube river system have emerged mainly focusing on environmental issues but also interested in navigation and water resource management. This has not made cooperation easier or more effective.

The main challenge to cooperation within the Danube waterway system, however, is arising from the merging of the Rhine and Danube regimes as a result of the operation of the Main-Danube Canal.

The Central Commission on Rhine Navigation (CCRN) was set up more than a century ago in 1868 following on from the convention signed in Mannheim. The Mannheim Convention established the principles of a navigation system on the Rhine that are still in force today. Participating states of the Mannheim Convention and the bodies of the CCRN with headquarters in Strasbourg are the Netherlands, Germany, Belgium, France and Switzerland. Austria is interested in joining the group.

Over the years, the CCRN has created a fully-fledged Regulation Act for inland navigation which comprises:

- Regulations for the operations of vessels navigating the Rhine which are supplemented by special technical rules
- Inspection Regulations for Vessels on the Rhine to ensure a unified mode of inspecting vessels and the award of a vessel's certificate for the Rhine
- Regulations pertaining to shipmaster's licences for Rhine navigation setting forth a uniform examination procedure for the acquisition of the shipmaster's licence
- Rules to apply in addition to the regulations mentioned above to all vessels carrying hazardous goods on the Rhine

The resolutions, rules and regulations drawn up by the bodies of the CCRN have the force of law for all vessels navigating the Rhine.

It appears that two issues would be of paramount importance for institutional cooperation in Danube-based navigation.

Firstly, the option of membership in both institutions as a precondition for harmonising the regulatory schemes for both river systems and/or a mutual acknowledgement of the other's regulations (and their implementation). The EU should play a decisive role in that process.

Secondly, the improvement of navigation conditions should be made the subject of a transnational investment programme, comprising all interested countries inside and outside the EU. Such a programme would boost national intentions with regard to infrastructure investment and make them more effective. It would also have a positive effect on private investment in inland navigation, which is urgently needed if the sector is to flourish.

The ability to draw up such a programme on a European scale would constitute nothing less than a test case for effective transnational cooperation and could well attain a significance extending far beyond inland navigation or even the transport sector as a whole.

References

Das technische Konzept des Main-Donau-Kanals, München 1992

Die Donau – Übersicht über den Schiffsverkehr auf der Donau (von Kelheim bis zur Mündung ins Schwarze Meer), Duisburg 1992

European Conference of Transport Ministers (ECMT) (1990) Inland waterway transport in ECMT countries to the year 2000 – A new dimension; Paris

Fischer MM, Rammer C (1993) Zur Entwicklung des grenzüberschreitenden Warenverkehrs im Einzugsbereich der Rhein-Main-Donau-Wasserstrasse. Mitteilungen der Österreichischen Geographischen Gesellschaft, 135. Jg. (Jahresband); Wien

Kienbaum and Partner (1996) Regional Study for the Danube Corridor Development Plan. Draft Final Report, vol I and III; Gummersbach

Österreichisches Institut für Raumplanung (1995) Beiträge zur Planung des Nationalpark Donau-Auen, Teilgutachten Schiffahrt (2 Bände); Wien

Planco (1995) Bewertung des Donauausbaues zwischen Straubing und Vilshofen, 1. Teilbericht; Essen

Programme of Development for the Infrastructure of Inland Navigation, Bureau for Research Organization and Development at the Institute for Transport Sciences, Ltd., Budapest 1994, for the Ministry of Transport, Communication and Water Management

Seitz, M (1997) Wasserstrasse Donau – Feldweg oder Highway. In: RAUM 26/97, Wien

Water Preservation Strategies and the Regional Plan for the Danube Delta Biosphere Reserve (DDBR)

Simona Pascariu
Urbanproiect Romania
Bucarest, Romania

1 Introduction

Water conservation strategies and the regional plan for the Danube Delta Biosphere Reserve (DDBR) offer a synthetic view of Romanian efforts to find the right solutions to two major problems:

- how to protect water resources, and
- how to set up a framework for regional planning and management of the DDBR.

Both aspects are deeply related. The Danube Delta is an important feature of the water resources of Romania. Concomitantly, DDBR represents a unique ecological site of major international importance as well as having a strong identity within the Black Sea Region and in Europe as a whole.

Given the general context (Figure 1) of viewing problems with a spatial impact as part of the process of integrating the CEE (central and eastern European) countries in the new greater Europe (as a natural consequence of geopolitical givens), the elaboration of the studies referred to addressed the following main issues:

- **European Strategy**
  the European Spatial Development Perspective (ESPD) and the fundamental and operational objectives of the ESPD, set out at Leipzig in 1994 for the territory of the Union, with impact at continental level;
- **Sustainability**
  for protection and sustainable development, considered as the correct middle route between the need for environmental regeneration and tendencies manifested at territorial level; it is the ideal path between the slowing down of the pace of development for the purpose of ecological regeneration and support for development within a monitored and regulated framework that squares the rhythms of enterprise with the need for ecological equilibrium;

- **Connections**
  the creation of major communication corridors, to establish a functional and efficient cross-European transport network; this network includes, among others, the navigable Rhine-Main-Danube route (the 7th Corridor, Crete, 1994) linking eastern and western Europe by water;

- **Dialogue**
  the establishment of an effective dialogue based on mutual exchanges of know-how and cooperation between the countries of the Black Sea Region; the main objective of this dialogue is the creation of a global strategy aiming to establish responsible policies with regard to natural assets and the human heritage; the strategy should also be directed towards the global needs of the common area of the neighbouring regions.

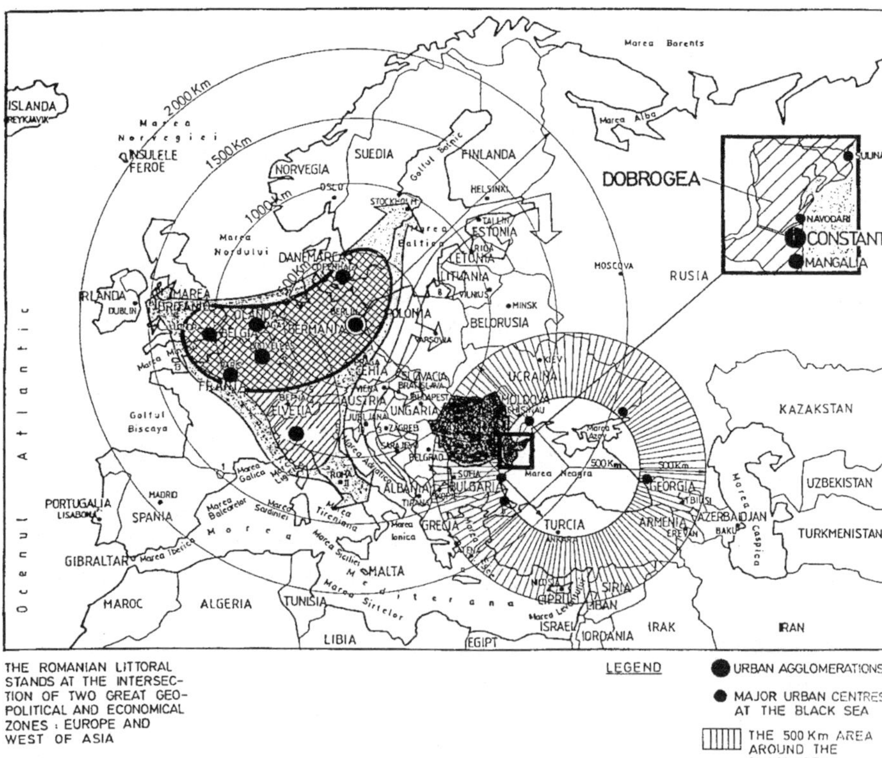

Figure 1. The Romanian Littoral in Europe

2 Water Conservation Strategies

2.1 General Considerations

In Romania, spatial organisation and regional planning are important tools for reducing regional disparities and the existing dysfunction at national level. At the same time, they have a crucial contribution to make towards the process of European integration.

The National Physical Development Plan (NPDP) represents a major means of dovetailiing sectoral and local policies with initiatives within the framework of spatial development. In the long run, the NPDP will give rise to global schemes and a global approach to regional planning aiding the spatial configuration of the national territory through:

- the establishment of long-term development programmes for the technical infrastructure, the management of vital resources, and the preservation of national assets and the cultural heritage in Romania;
- integration within the development concepts of the European and international programmes that Romania is already part of due to its geopolitical and strategic position and overall potential (i.e. the Rhine-Main-Danube navigable channel, the Black Sea Region, the Danube Delta etc.).

NPDP outlines in a global, long-term manner the general planning strategy for the Romanian national territory in the following areas:

1st Section of the NPDP: main communications and transport infrastructure by land, water, and air;

2nd Section of the NPDP: the sustainable management of water resources whilst meeting required standards;

3rd Section of the NPDP: conservation of the nationally important natural and built heritage;

4th Section of the NPDP: the sound development of a balanced and efficient human settlement network, based on the socio-economic and cultural-historical calling of communities to support sustainable development.

2.2 The Objectives of the 2nd Section of NPDP: Water

In accordance with the general principles adumbrated above, the main objectives of the **2nd Section of the NPDP - Water**, are as follows:

I. efficient use of water resources in every socio-economic sector;

II. to increase water resources by building up hydrological reserves and improving distribution in the territory by means of transfer pipelines and feeder systems from richer areas to poorer ones;

III. conservation and improvement of the quality of both surface and underground waters;
IV. to enhance levels of hygiene and sanitation and, by extension, the quality of life in human settlements by providing drinking water for domestic consumption to the required standards;
V. rehabilitation and development of the hydrological systems for localities, in accordance with the requirements of related development programmes;
VI. popular consciousness-raising concerning the importance of public water utilities and direct involvement of the population in campaigns to reduce and eliminate pollution of water resources.

2.3 Water Resources

Data for Romanian water resources, with regard to the main hydrological basins, are currently as followings:

Water resources	Theoretical potential (billion cu.m/year)	Technically available potential (billion cu.m/year)
SURFACE WATERS:		
• Rivers, interior lakes	41.0	16.9
• Danube	85.0	20.1
UNDERGROUND WATERS	9.0	5.7
TOTAL	135.0	42.7

Black Sea resources are extremely important, but cannot be taken into consideration due to the technical and economical difficulties relating to desalinisation.

Specific water resource availability for the Romanian population totals:

- 1,625 cu. m per inhabitant per year from surface waters, and
- 250 cu. m per inhabitant per year from underground waters.

Overall specific water resource availability stands at 1,875 cu. m per inhabitant per year. Romania accordingly has one of the lowest levels of water resources of any European country.

For the time being, the technically available potential cannot be effectively used unless some important investments are effected. These relate to hydrological works in major basins that are vital bearing in mind that:

- the Danube river, which is the most important water resource in Romania, cannot be effectively utilised for social and economic development due to its marginal geographic position;
- the domestic river basins have an unequal distribution, leading to the situation where large areas suffer from scarcity of water resources;

- the alternation of dry and rainy years and the high variations of small and medium size river flows often generate floods in different regions of the country;
- the pollution of some domestic rivers strongly reduces the availability of water resources.

2.4 Problems

The specific problems of water resources and of their utilisation in Romania are:

- the continuous increase in water consumption, mainly for the needs of the technical infrastructure of communities;
- downturn in investment rhythms with regard to public works projects directed at increasing water resource availability (e.g. water quality enhancement, water capacity enhancement, flood prevention) induced by the scarcity of financial resources;
- decrease in the quality of water resources, due mainly to their inadequate utilisation[1] and protection;
- the perceptible degradation of underground water resources in recent years due to expansion of the affected areas and the intensified impact of polluting factors;
- increase in the number of pollution incidents arising from disrespect for technological and production regulations and damage caused by diverse violations of existing legislation (mostly by the petrochemical refineries).

2.5 Water Conservation Strategy

The water management sector is a nationally crucial technical infrastructure asset. It has the task of providing, now and in the long run, the necessary volumes of water for both private and industrial consumption and also of taking protective action against the potentially destructive effects of natural water hazards.

Water conservation strategy has been drawn up in accordance with the above and involves rational management and protection of water resources, avoidance of excessive consumption and pollution, and observation of the principles of environmental protection and sustainable development.

---

[1] Of a total of 20,442 km of waterway subjected to quality analysis, the breakdown is as follows:
11,105 km (54.3 %) - 1st Category waters;
4,027 km (19.7 %) - 2nd Category waters;
2,218 km (10.8 %) - 3rd Category waters;
3,092 km (15.2 %) - 4th Category waters (degraded)
Drinkable water resources are represented by the 1st Category.

2.6 Implementation

The national strategy for the conservation of water resources, the **2nd Section of NPDP: Water**, was adopted by the Romanian Parliament on 9 September 1996 as a legal directive in the field. The Act provides for a number of initiatives to increase and conserve water resource potential in the long term.

The most significant of these are:

a) to complete work on water reservoirs wherever execution has already commenced, on the basis of specific cost-benefit studies;
b) to erect the feeder and transfer pipelines necessary for the supply of drinking and industrial water;
c) to conserve and improve the quality of surface and underground waters through modernisation and use of new, modern technologies for water purification facilities and by building new such facilities wherever this is necessary;
d) to complete already initiated capital expenditure on water supply and sewerage systems;
e) to upgrade drinking water treatment plants and provide the necessary buildings and installations for water storage;
f) to reduce spillage along supply pipeline networks by using materials with a higher degree of resistance to corrosion and by installing meters to record the quantity of water supplied;
g) to increase the efficiency of water recycling during the production process up to the limits of existing installed capacities, or to provide new recycling facilities.

3 The Regional Plan for the Danube Delta Biosphere Reserve

3.1 General Data

At the regional level of the counties[2] or other aggregated areas (several communes or counties), the Romanian spatial development policies aim:

- to implement regional physical plans;
- to facilitate the elaboration of a strategic management concept for spatial development;
- to create an integrated framework for public investment programmes; to administrate and monitor the sectoral funds for regional equipment, the technical in-

[2] The administrative levels in Romania are the Communes and Towns (local) and the counties or *JUDETS* (regional).

frastructure at regional and local level (roads or other services to a group of human settlements), and housing and social facilities for culture, health, education, sport and leisure.

The most important inhabited natural areas in Romania for which regional physical development plans have been elaborated are the western Carpathian Mountains (Apuseni) and the Danube Delta.

Since 1990, the Romanian Parliament has adopted a number of legislative measures relating to the *Danube Delta* and the *Razim-Sinoe Lagoon System*. Among the most important decisions are the following two:

- Act 82/1993 relating to the setting up of the Danube Delta Biosphere Reserve (DDBR) and
- Governmental ruling 248/1994 setting out detailed measures to aid implementation of Act 82 (DDBR).

At the same time, the Reserve is organised in accordance with the objectives of the most representative reserves from the network for natural ecosystems, supported by Project 8 of the international "Man and the Biosphere" programme (MAB-UNESCO), final draft 22/1974.

The DDBR area is situated in the south-eastern part of Romania and includes several specific physical and geographical features. Among the most important are the Danube Delta proper, the salted area of Murighiol-Plopu, the Razim-Sinoe Lagoon System, the inland maritime waters and the territorial sea area, up to the 20m-depth isoline.

The surface of the DDBR covers 5,800 sq. km and comprises:

- *506 sq. km of totally protected zones,*
- *2,333 sq. km of buffer zones,*
- *3,061 sq. km of economic zones.*

The human population of DDBR stood at 14,315 as of 1 January 1996.
The settlements network includes one town (Sulina) and 24 rural localities.

The Reserve territory represents an ecological site unique both in Europe and among the world's delta ecosystems. During the last decades the Danube Delta area and its contiguous areas have been subjected to a number of important ecological interventions, with evident negative effects. The upshot was global ecological decline in the form of:

- reduction of bio-diversity within the Delta system;
- degradation of the working and living conditions of inhabitants within the Delta area;
- high pollution of the Danube waters.

These circumstances necessitated setting in train a radical shake-up of the organisational and management systems in the area. *The main objectives of the proposed strategy are the regeneration of fundamental ecological processes and the improvement of living conditions.*

The Regional Plan is an instrument which aims to unify the different policies for the Delta Reserve, applying them to the surrounding areas too. It is considered as an enabling tool for the accomplishment of global and sectoral public policies and for the setting up of a conceptual framework for spatial development. This global planning strategy has been elaborated in accordance with the latest national and international principles and criteria for world biosphere reserves.

3.2 Considerations about the Impact of Human Activities on the Delta and Littoral Ecosystems

The natural and economic decline manifested during the past three decades in almost every field of activity within the DDBR area is grounded in several factors, among the most important being those relating to human social and economic activities. The most significant factors to have been identified are:

- the impact on the natural hydrological cycle of the erection of a significant number of canals, dams and polders and of river re-routing measures;
- the over-exploitation of biological resources (industrial over-fishing, poaching, etc.);
- the degradation of natural landscapes through environmentally deleterious tourism;
- the total transformation of large areas of wet-lands into purely agricultural areas;
- the increase in the ecological risk potential due to uncontrolled shipping development;
- the rising percentage of toxic chemical substances entering soil and water due to the intensive use of fertilisers for agriculture and forestry;
- the lack of social protection policies for inhabitants of the Danube Delta, leading to rising conflict zones as regards the environment;
- the non-cost-effective management of reed production, a basic traditional resource in the area.

3.3 The General DDBR Planning Concept

The DDBR planning and development strategy incorporates a number of principles, elements and data from international documents produced by specialist institutions as well as drawing on Romanian legislation for the area and on the Physical Regional Plans for Tulcea and Constanta counties.

In accordance with these elements, the main goals and functions of the Regional Plan for the DDBR focus on the following issues: *the sustainable conser-*

vation of natural resources for the benefit of coming generations; the preservation of the traditional use of the land; the improvement of natural resource management for the benefit of local populations, the avoidance of environmental degradation; international cooperation in solving the problems relating to the safeguarding and management of natural resources; the provision of scientific information for the world biological reserves network.

The DDBR statutes, enshrined as Law 82/1993 and GR. 248/27.05.1994, offers:

The local communities:	utilisation of natural resources with the help of long term protection and conservation measures; job creation in the field of public safety and research; a healthier and cleaner environment; real participation in the decision making process concerning the utilisation of natural resources;
The resource managers:	the opportunity to rehabilitate and prevent the consequences of inadequate use of resources;
The world community:	information about options available for improving natural resource utilisation.

The above mentioned legislation, also stresses a number of concrete and coherent principles relating to the organisational structures and functions of DDBR. For the planning and development process of the Reserve, the most important of these are:

A. The ownership structures of real estate. This stipulates diverse methods of administration and exploitation and leads to separate managerial procedures for:
   - *public national interest estates, overseen by the Danube Delta Biosphere Reserve Administration (DDBRA);*
   - *private properties of physical and juridical persons or of local authorities, public or private estates.*

B. Separate administrative rules for natural resources than for residential development so as to protect and conserve them, with the goal of integrating and organising the various activities to accord with the sustainable natural yield potential of the Delta environment.

To conclude, the Regional Plan for the DDBR lists severely restricted areas where the free access of people and the practising of some activities are totally forbidden.

On the other hand, the plan identifies the areas where the population is allowed to carry on traditional economic activities that do not harm the environment.

3.4 Regional Planning Framework for the DDBR

To formulate the planning and development proposals for DDBR, it is necessary to take into account a number of spatial elements of greater regional significance, such as:

- firstly, the geography of the Reserve, located as it is in an area straddling the Romanian and Ukrainian borders; this cross-border location calls for a coordinated bilateral spatial policy;
- secondly, there are the role and functions of the river Danube as a major trans-European navigable route; this provides settlements within the DDBR with enormous potential for the development of diverse social and economic activities such as industry, shipping, tourism, trade etc. Questions of water quality and the protection of ecosystems demand a high degree of cooperation between all the riverine countries so as to enable practical measures to be adopted to reduce and eliminate pollution of the Danube brought about by a wide range of human activities, including transportation;
- thirdly, the DDBR territory is a component of the Black Sea Region, which is governed by the *Black Sea Economic Cooperation Treaty (BSEC)* signed by 11 riverine and non riverine countries. *BSEC* is having a major spatial impact on development of the littoral zones and regional planning for the countries involved.

Economic integration and technical infrastructure programmes for the BSEC countries embrace both the DDBR and its surrounding area, justifying the correlation between these programmes and the objectives of ecological protection for this area.

Besides macro-regional spatial entities, thought must also be given to the national regional context and related local issues. Bordering the DDBR, the terrestrial areas of the Tulcea and Constanta counties have a major impact. Within their bounds, there are many interrelated activities and processes of an economic, social, and transport-related nature which likewise necessitate special coordinated and integrated development policies for economic and social activities, human settlements, and infrastructure.

3.5 DDBR Regeneration Programme

Due to the complexity of the DDBR's problems, the Romanian Government has decided to elaborate a *Special Development Programme to Support Tulcea County and the DDBR.* Adopted on 2 April 1996, the programme foresees economic and social policies and initiatives for a comprehensive regeneration of the area under *Ministry for Public Works and Regional Planning (MPWRP)* coordination.

This regional development programme will have a major economic impact, involving serious efforts on the part of government in cooperation with local authorities and other important players, the private sector for instance. The programme is made up of:

I. the Physical Regional Development Plan for Tulcea county;
II. the Physical Regional Development Plan for the DDBR area;
III. the socio-economic development strategy for Tulcea county 1995–2004 and
IV. the priority objectives for Tulcea county, elaborated by the Tulcea County Council.

The main objectives and guidelines of the programme are set out in governmental and local authority rulings on the following major aspects:

- *the sustainable utilisation of natural resources and of economic assets* through the reshaping, modernisation and diversification of economic activities, such as agriculture, fishing and fishery, industry, environmentally sensitive tourism, trade and transportation;
- *the stabilisation of the DDBR population* by creating adequate living conditions, social facilities and economic incentives;
- *the protection of the natural and the built heritage* by devising special protection policies for the natural ecosystem of the DDBR;
- *the generation of attractive economic structures* for foreign investors to reflect the strategic position of the DDBR on the Danube and within the Black Sea Region.

The programme provides for sectoral activities in four stages of implementation, viz. by *1996, 1997–2000, 2001–2004 and 2004 onwards*. It proposes detailed initiatives as well as investment objectives.

The most important measures and initiatives foreseen by the programme relate to:

I. the rehabilitation of natural terrestrial and aquatic economic potential through the regeneration of biological resources and the implementation of targeted and effective methods for the protection of the natural and cultural heritage;
II. social measures, through support of economic sectors so as to create secure employment structures and stable revenues for the local population, as well as through rehabilitation and regeneration of less favoured or declining areas, social policies suited to the existing demographic structure, and improvements to the technical infrastructure and social facilities;
III. the fulfilment of the scientific potential of DDBR through astute management;
IV. the promotion of scientifically organised ecosystems;
V. the physical integration of the DDBR at local, regional and international level by means of a modern and reliable technical infrastructure;
VI. the preservation of the cultural identity of the DDBR through the promotion of its traditional activities, through environmentally sensitive tourism, and through other PR and information-related activities inside and outside the DDBR.

4 Development Trends and Specific Policies for the Romanian Littoral Zone (RLZ)

4.1 Synopsis

The study for the Romanian Littoral Zone integrates water preservation strategies with the objectives and principles of regional physical planning. The zone has been delineated using a multi-criteria approach and includes both areas of the DDBR as well as urban and rural and areas of touring interest.

In line with European principles in the field, the study aims to propose a specific strategy and management for the RLZ based on partnership between local public administrations and the public and private sectors. The proposed strategy, policies and programmes also aim to involve the local communities and public opinion in general.

The strategy is composed of general objectives and specific programmes to be adopted for implementation at various administrative levels. The main constituents of the strategy are set out in the following boxes.

• **STRATEGIC OBJECTIVES**	
– *for the medium term*	reducing and where possible eliminating the causes of pollution and improving living conditions in residential areas;
– *for the long term*	ecological regeneration and sustainable development of the whole area.
• **MAIN POLICIES**	
– development policies	relating to the technical infrastructure at territory, settlement, economic zone and social institution level;
– protection policies	relating to the natural and the built environments and to cultural and spiritual values;
– promotional policies	in fields like education, public information, settlements and the legislative system.
• **LEVELS OF AUTHORITY**	
– local	development (economic, financial, urban, etc.)
– *judet*	protection (natural and built environment, heritage, land, etc.)
– government	promotion (education, information, legislation, etc.)
• **SPECIFIC PROGRAMMES**	
– reconstruction programmes	• development of the structure and facilities of the territory • development of public facilities and equipment in settlements • improvement of living conditions

– support programmes	• protection and rehabilitation of the natural environment • protection and regeneration of the built environment • development of the service sector • public information and education
– framework programmes	• drawing up of integrated development plans • establishment of a legislative and institutional framework in the RLZ • setting up of an integrated operational and monitoring system for the RLZ
– integration programmes in regional, European and international programmes for the Black Sea Region	

4.2 Challenges and Specific Problems for the Littoral Sub-Zones

Geographical characteristics, historical development, principal communication links and the distribution of natural resources have given rise to an uneven distribution of population and economic activities within the RLZ *(Figure 2)*. This process of differentiation permits a division of the territory along typological and functional lines into 5 *SUB-ZONES (SZ)*, which can be regarded as *specific territorial units*. These units are summarised in the following boxes.

SZ 1	**MARITIME DELTA ZONE** (subdivision of the DDBR)
	■ surface area: 1,500 sq. km; ■ human population: 11,000 (2 % of the total population of the RLZ); ■ problems and policies: SZ 1 has problems relating to the physical and morphological adulteration (degradation) of its natural environment; development of the area is governed by Act 82/1993 (7 December) for the DDBR. Strong government support is needed for environmental protection and the provision of physical infrastructure, technical equipment and housing.
SZ 2	**LAGOONARY SYSTEM ZONE** (partially included in the DDBR)
	■ surface area: 2,000 sq. km (of which 40 % is water); ■ human population: 30,000 (5 % of the total population of the RLZ); ■ problems and policies: SZ 2 is confronted by problems relating to chemical, physical and sociological degradation with important economic implications. This sub-zone needs specific incentives for the development of tourism and of its traditional economic activities, plus strong support for rehabilitation from the government. Both governmental and local support are needed.
SZ 3	**CENTRAL ZONE** (between the two branches of the Danube - Black Sea Channel)
	■ surface area: 965 sq. km ■ human population: 500,000 (77 % of the total population of the RLZ) ■ problems and policies: this zone is characterised by strong urban concentration and by important currents of enterprise (economic, public equipment and utilities, etc.) across the entire area; significant levels of pollution are another main feature The zone needs clear regulatory development schemes, notably as regards land use and specific financial policies at local authority level.
SZ 4	**HEAVILY-TOURED LITTORAL ZONE** (south-eastern coastal area)
	■ surface area: 350 sq. km ■ human population: 75,000 (12 % of the total population of the RLZ) ■ problems and policies: the heavily-toured littoral area needs specific policies and actions directed at the protection and efficient use of tourism assets and natural potential; international environmental standards also need to be met within this area. It is important that the *judet* and local authorities provide enough support to solve problems that exist in the fields of housing, equipment, and pubic utilities.
SZ 5	**INTERMEDIATE ZONE** (includes the south-western zone, adjacent to the tourist area)
	■ surface area: 865 sq. km ■ human population: 27,000 (2 % of the total population of the RLZ) ■ problems and policies: SZ 5 is a rural territory with significant development potential (agricultural land, human and natural resources), needs careful and appropriate development policies. The area needs support from both local and central level, integrated development plans, and prioritisation.

In February 1997, the *judet* councils started the process of approving the strategy for the Romanian Littoral Zone set out above.

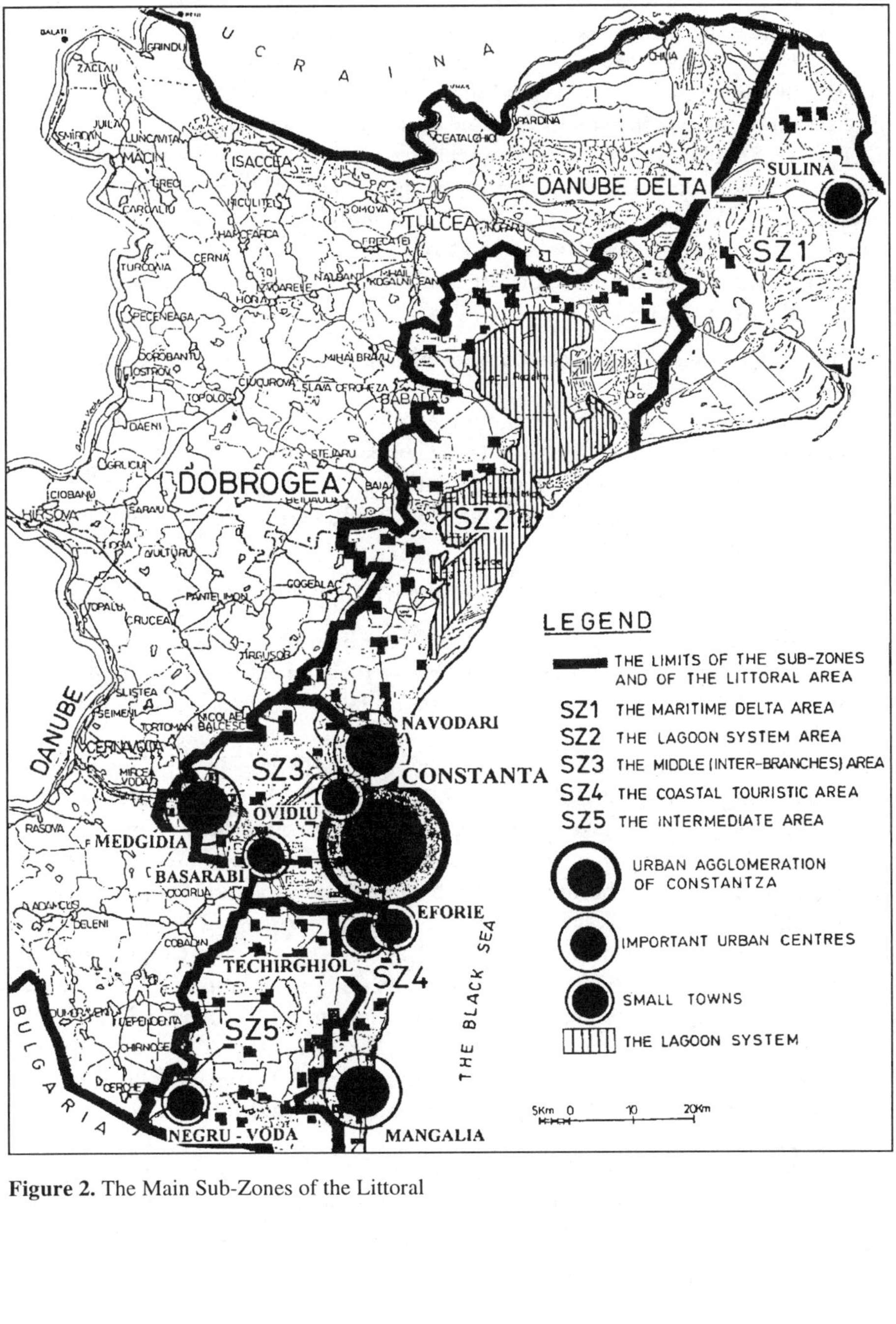

Figure 2. The Main Sub-Zones of the Littoral

5 Conclusions

Water conservation strategy in Romania has been elaborated with a view to supplying water, now and in the future, for both private and industrial needs by rationally managing water resources and by safeguarding against over-consumption and pollution, in accordance with the principles of environmental protection and sustainable development. One of its main objectives is the conservation and improvement of the qualities of surface and underground waters.

The DDBR planning and development strategy incorporates a number of principles, elements and data from international documents produced by specialist institutions as well as drawing on Romanian legislation for the area and on the Physical Regional Plans for Tulcea and Constanta counties.

In accordance with these factors, the main tasks and goals of the Regional Plan for the DDBR are as follow:

- sustainable conservation of natural resources for the benefit of coming generations;
- preservation of the traditional use of the land;
- improvement of natural resource management for the benefit of local populations, avoidance of environmental degradation;
- international cooperation in solving the problems relating to protection and natural resource management;
- provision of scientific information for the world biological reserves network.

The experience of the past seven years in the field of environmental protection and regional planning underlines the necessity for a far more closely-knit coordination of individual projects, policies and initiatives at national and macro regional levels.

Water conservation and management of the DDBR are, in the final analysis, more than simple national challenges. They require integrated strategies and policies within the larger context of the Black Sea riverine countries.

The first conclusion, then, is that neither water resource conservation nor regional planning within a border area can be a problem of national concern only.

A good example of this is the River Cris Basin Management Programme in the west of the country that forms part of a PHARE Cross-border Cooperation Programme between Romania and Hungary initiated in the autumn of 1996. But there are also other bigger regional programmes such as the Danube, Black Sea, or Baltic programmes.

The second conclusion is that everyone living in this region, will be affected in the long run by decisions taken today. It is thus important for the decisions to be the right ones.

Finally, given its human and natural potential, the transport and communications options it provides, and the bridge it forms with Asia and northern Africa, the importance of the Black Sea Region in Europe needs to be underlined.

We must play our part in meeting the challenges posed with regard to natural resources in general and water resources in particular by establishing an integrated zonal management system, exploiting the opportunities offered by the various specialist programmes with international backing, and promoting coordinated action within the Black Sea Region.

VASAB 2010
Transnational Cooperation in the Spatial Development of the Baltic Sea Region

Jacek Zaucha
University of Gdansk and VASAB 2010 Secretariat
Gdansk, Poland

The Origins of VASAB 2010

International cooperation in the field of spatial planning started in the Baltic Sea Region (BSR) in 1992. On 21 August 1992, ministers responsible for spatial planning and development in countries of the Baltic region met in Karlskrona in Sweden to discuss the prospects for future spatial development in the region. They decided to initiate permanent cooperation in BSR spatial planning at governmental level by jointly preparing a document on spatial development prospects for the BSR. The programme was named "Vision and Strategies around the Baltic Sea 2010" (VASAB 2010). A body known as the Group of Focal Points was set up to conduct the work. The Group comprises representatives (usually heads of spatial planning departments in the relevant ministries) from Belarus, Denmark, Estonia, Finland, Germany (at both federal and regional level – Mecklenburg-Vorpommern and Schleswig-Holstein), Latvia, Lithuania, Norway, Poland, Russia (Kaliningrad, Karelia, St. Petersburg regions), and Sweden. The Group of Focal Points is backed up by a secretariat.

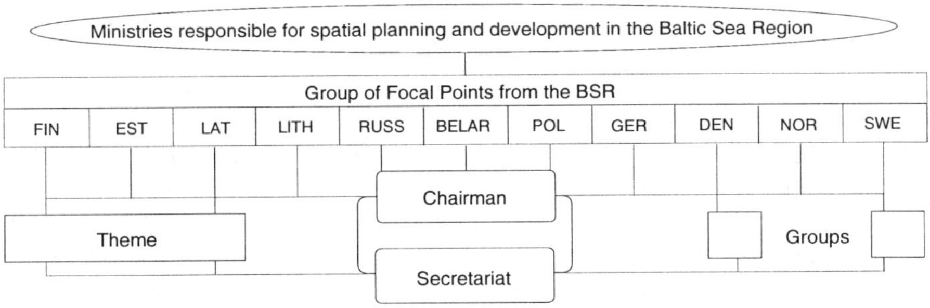

VASAB 2010 Chairpersons:
1. Mr Jan Kuligowski (Poland) 1992–1993
2. Ms Sirkka Hautojärvi (Finland) 1994
3. Mr Jacek Szlachta (Poland) 1995 (first half)
4. Mr Bo Kjellén (Sweden) 1995–1996
5. Ms Dzintra Upmace (Latvia) since October 1996

VASAB 2010 secretariat:
1. Sweden (Karlskrona) 1992–1995
2. Poland (Gdansk) since 1996

Figure 1. Cooperation in the Baltic Sea Region

The following objectives have been formulated for VASAB 2010:

- to support the development of cooperative networks within the BSR,
- to provide a practical forum for the transfer of powers to countries making the transition to democratic market economies,
- to improve the level of information in BSR countries on current trends and problems in other countries of the region,
- to evaluate major infrastructure projects and secure international financing,
- to promote spatial planning in the participating countries and regions.

Strategic Concept for Spatial Development in the BSR

Since its very beginning VASAB 2010 has focused its efforts on strategic planning, as is witnessed by the common strategy (vision) for spatial development in the area adopted at a conference in Tallinn (1994) by all ministers for spatial planning and development from the Baltic Sea countries. The decision to approve the 'vision' was taken unanimously by ministers, indicating the strength of will amongst all countries bordering the Baltic to contribute to the common prosperity of the region through spatial planning. The BSR is thus one of a select group of European regions having an advanced common strategy on spatial development. The reasons for the BSR having come such a long way in joint work on strategic issues are numerous and interrelated:

1. The Baltic has historically enjoyed a very strong sense of identity. The Hanseatic League is one example, but relations between the Baltic States and the Scandinavian countries have also always been very strong, even when political circumstances have been less propitious.
2. The spatial cohesion of the BSR has been jeopardised by the political realities of the past fifty years. The iron curtain dividing the region stifled harmonious spatial development for 45 years. At the beginning of the Nineties it was obvious that only common efforts could reverse these negative trends and restore more positive patterns from the past.
3. However, the political and economic transformation of the BSR in the early 90s also created fresh obstacles to harmonious spatial development in the BSR. Restored borders sometimes dramatically changed the economic and social relations that had taken shape. The Kaliningrad region, for instance, has been turned into the largest army camp in Europe. Multilateral cooperation at regional and national level thus suddenly appeared to be one of the best remedies against existing and newly emerging barriers to the development of the region.
4. When it set out on the road of multilateral and transnational cooperation in spatial development, the BSR was in the fortunate position of being able to avail itself of Scandinavian and German funding and knowledge. In this context, experience gained by the EU (Europe 2000 and Europe 2000+) was very important.

5. In the Nineties, European decision-makers began to regard spatial cohesion as an important precondition for the socio-economic development of Europe in a global context[1]. The goals and philosophy enshrined in VASAB 2010, with their focus on long-term "soft" development measures integrating different sectoral policies, were of a piece with the EU approach. In the long run it is expected that the VASAB 2010 strategy will attract new investors and facilitate socio-economic development of the region.
6. Since 1996, when Sweden and Finland joined the EU, the Baltic Sea Region has also become more central to the affairs of the European Union. This has resulted in a more active attitude towards the BSR on the part of the European Commission.
7. The Baltic Sea Region has known positive developments in the sphere of multilateral cooperation, the prime example being the Scandinavian countries. There is a strong feeling among these countries that such cooperation can create an important synergistic effect
8. Almost all the BSR transition countries have been united by an overriding wish to become as integrated in western Europe as they once were before. Cooperation within the framework of VASAB 2010 has played an important part in the run-in to deeper integration.[2] This is because the main objective of VASAB 2010 cooperation has been promotion of regionally balanced and sustainable (economic, environmental and social) development and consolidation of the spatial cohesion of the BSR. This is in line with the main objectives of the *Treaty on European Union* (Title I, Article B and Title XIV, Article 130a) and *Principles for a European Spatial Development Policy* [3].

The Vision

Four basic values, relating to improvements in the quality of life, constitute the "heart" of the vision:

- *development*, going beyond economic prosperity and growth,
- economic, social and environmental *sustainability*, since development must not be to the detriment of future generations,
- *freedom*, in the sense that it implies the possibility to choose in accordance with individual/regional preferences,

---

1 Private investors need firm and stable guidelines on the intended future development of the infrastructure, cities, protected areas and other spatial components of the region. Developers expect regional and national authorities to update their plans and concepts in advance to create a stable business environment. Therefore spatial planning creates important preconditions for harmonious socio-economic development of the region.

2 Spatial planning encompassing the whole territory of the BSR is preparing the Associated Countries for future integration in the European Union.

3 Adopted by the Informal Council of Ministers responsible for Spatial Planning and Development in the Member States of the European Union at Leipzig in September 1994.

- *solidarity*, the characteristic of a caring society that shares the benefits of development.

In the vision, spatial structures are divided into three basic elements: the system of cities and urban settlements ("pearls"), the interlinking infrastructure networks ("strings"), and selected types of land use ("patches") in non-urban areas. They are promoted by "the system", i.e. planning institutions, rules and procedures. The report concludes with recommendations on how to achieve the intended vision.

Implementation of the VASAB 2010 Strategy

After the Third Ministerial Conference in 1994, VASAB 2010 set about implementing the strategy. This effort had a multidimensional character and involved, on the one hand, generating and effectuating spatial planning projects and, on the other, cooperating with other international bodies with a view to encouraging them to become involved in implementing a common strategy. It was obvious for VASAB 2010 leaders that only joint action involving different levels of public administration (local, regional, national) can ensure proper implementation of the VASAB 2010 strategy. Therefore from the very beginning VASAB 2010 worked together with all the Baltic regional and national networks. The VASAB 2010 strategy was sanctioned by the conference of regions within the BSR (Subregions Conference) – the main partner of national level involved in its implementation.

Pilot Projects and Joint Action

In 1994 the Focal Point Group became the VASAB 2010 steering committee and was henceforth known as the Committee on Spatial Development of the Baltic Sea Region (CSD/BSR).

First a list of pilot projects and joint initiatives was agreed upon by the CSD/BSR as a means of speeding up the implementation process by concentrating on the most important aspects of long-term spatial development in the BSR.

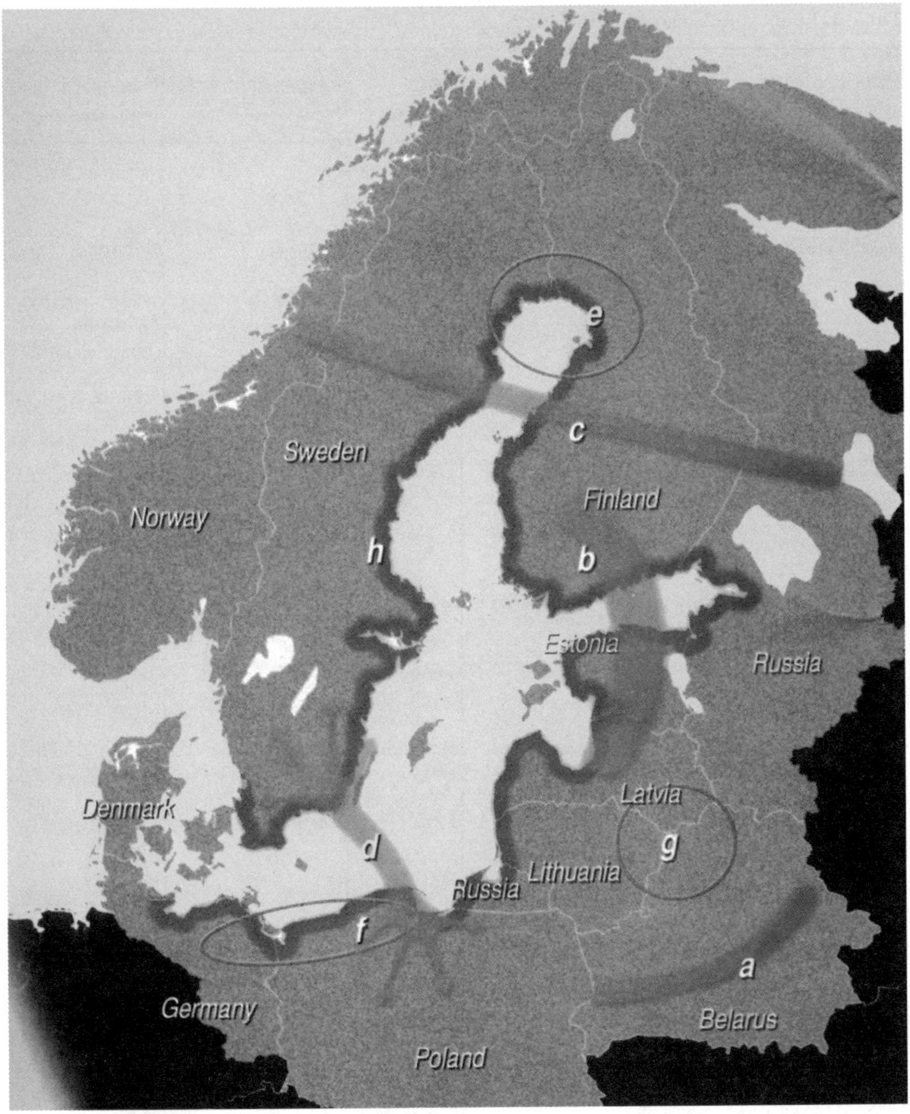

a Orsha–Minsk–Brest zone
b THTR Zone (Tampere–Helsinki–Tallinn–Riga)
c Karelia–Atlantic (Petrozavodsk–Vaasa–Umeå)
d TEM/TER (South East Baltic cooperation)
e Arch of Bothnia (Luleå–Haparanda–Tornio–Kemi–Oulu city networking)
f Project on tourism development of German-Polish border
g Project of transborder cooperation between Latvia, Lithuania and Belarus
h Common recommendations for spatial planning of the coastal zone in the Baltic region

Figure 2. Pilot Projects and Joint Initiatives
Source: Zaucha 1997.

Table 1. Pilot Projects and Joint Initiatives

Pilot Projects and Joint Initiatives	Time horizon	Responsibility	Other participating countries
1. Transport corridors as zones of accelerated development			
• Orsha–Minsk–Brest	1996–97	Germany	Belarus
• THTR (Tampere–Helsinki–Tallinn–Riga)	1995–97	Finland	Latvia, Estonia Lithuania (observer)
• Karelia–Atlantic (Petrozavodsk–Vaasa-Umeå)	1995–97	Finland	Karelia, Sweden
• TEM (Katowice–Gdynia–Karlskrona)	1996–97	Poland	Sweden
2. Urban system			
• Bothnia Arch – Municipality cooperation in the northern part of the BSR	ongoing	Sweden/Finland	
3. Specific areas			
• Development of tourism in German–Polish border region	concluded 1995	Germany/Poland	Kaliningrad as an observer
• Transborder cooperation between Latvia, Lithuania and Belarus	1995–97	Germany	Latvia, Lithuania, Belarus
4. Exchange of knowledge			
• Compendium of Planning Legislation (linked with EU Compendium)	1996–97	Estonia	all BSR countries
• Seminar on National Strategic Planning Concepts	Stockholm, September 6–8, 1995	Sweden	all BSR countries
5. Planning system			
• Common recommendations for spatial planning of the coastal zone in the Baltic region	concluded 1996	Denmark	all BSR countries

Of the pilot projects, those addressing the use of spatial planning to transform transport corridors into zones of accelerated development were most advanced. Much interest was aroused by the plan to create a network of marinas on the southern and the eastern coasts of the Baltic Region as well as by transborder co-operation schemes with strong spatial elements (Poland/Germany/Lithuania/ Latvia/Belarus).

Secondly, efforts aimed at securing adequate resources for the implementation of the above-mentioned projects were set in train by the CSD/BSR at project conception. This led in 1996 to INTERREG II C being formulated to take account of the needs of multilateral spatial planning for the BSR.

The Fourth Ministerial Conference

The Conference held in Stockholm (October 1996) and chaired by Germany approved the priorities established by the Committee on Spatial Development of the Baltic Sea Region. The Ministers also adopted concrete findings from the initial VASAB 2010 implementation phase such as *Common recommendations for spatial planning of the coastal zone in the Baltic region*, and proposed project-oriented application of these recommendations. The Conference discussed the issue of INTERREG II C, PHARE and TACIS integration and approved the VASAB 2010 plan of action for the years ahead dubbed "From Vision to Action". The "Stockholm Declaration" provided strong political guidance for future work with VASAB 2010. In this document, the Committee on Spatial Development of the Baltic Sea Region was urged to concentrate on the following initiatives:

- setting up a programme on sustainable settlement and urban network development ("pearls");
- combining the development of a European Transport Network with sustainable regional development, i.e. with reference to the following corridors ("strings"): Via Baltica, Via Hanseatica, TEM/TER, Atlantic–Carelia, southern Finland–St. Petersburg–Moscow, Berlin to Moscow, and from Copenhagen, Ystad, and Malmoe to Szczecin and on southwards;
- establishing a set of pilot projects to oversee sustainable development of specific areas ("patches"), i.e. cross-border areas, islands, lakelands, cultural landscapes, rural and forest areas, coastal zones, nature conservation areas;
- creating a network of marinas taking account of their regional and environmental impact;
- building up a spatial development monitoring system for the BSR;
- organising regional conferences for the presentation of key projects and a permanent project management system for the implementation of VASAB 2010.

The Agenda for the Future Work of VASAB 2010

BSR Urban Network

A balanced settlement structure is essential for sustainable spatial development of urban and rural areas. Key issues in supporting the development of rural areas are decentralisation of settlement, business and administrative activities. After the 3rd Ministerial Conference, a pilot project on city cooperation (Luleå–Haparanda–Tornio–Kemi–Oulu) was launched to demonstrate the benefits resulting from co-operation between cities in servicing their hinterlands, functional division of labour, exchange of experience, and harmonisation of the main aspects of their development. VASAB 2010 joint action on urban networks in the BSR aims to design an operational strategy for achieving a balanced system of settlement around the Baltic, and this will require the involvement of various institutions at local level as well as at subnational, national, and international levels.

It has been proposed that joint VASAB 2010 action on the Baltic urban network be initiated in 1997. The detailed terms of reference of VASAB 2010 work in this field are under preparation – with Denmark being appointed as the chair country.

Using Transport Corridors for Sustainable Regional Development

Development of the transport network in the BSR is within the remit of the Conference of the Baltic Ministers of Transport complemented by the G 24. The goal of VASAB 2010 is to provide a spatial framework for the development of the BSR transportation network. In particular, the focus of spatial planning is on the impact of infrastructural development on the socio-economic development of the surrounding (sub)region and the natural environment of the adjacent area. This is to optimise sectoral investment in subnational development as a means of shifting attention from pure road building to analysis of settlement structures, the green areas system, and alternative, intermodal forms of transport with a view to harmonising the socio-economic development of areas adjacent to communication lines and hence to conserving their natural resources and landscape values. Spatial planning also helps bridge the interests of subnational, local and national players from different sectors with regard to transport corridors.

Four pilot projects involving transport corridors accorded accelerated development status (Tampere–Helsinki–Tallinn–Riga, TEM/TER, Brest–Minsk–Orsha/C 9, Karelia–Atlantic *Blue Way* corridor) have been launched so far to investigate how transport corridors can be utilised in the interests of sustainable regional (subnational) development. The objectives of the projects are given below:

Table 2. Objectives of VASAB 2010 projects

Objective	Projects
Environmentally sound economic development (in the sense of Agenda 21) of the zone under consideration	T-H-T-R, TEM/TER, C 9
with more specific subtasks such as: • preparation of a set of thematic maps on the structure of the zone • establishing an information system on existing development potential for investors in the corridor	C 9
Conservation of nature and landscape areas of special interest	T-H-T-R, C 9, TEM/TER, Blue Way,
Development of tourism	T-H-T-R, Blue Way, TEM/TER, C 9
Spatially balanced development both at national and subnational level	T-H-T-R, TEM/TER, C 9
Development which respects cultural aspects of human life and local traditions	T-H-T-R, TEM/TER
Aspects of quality of life such as education and employment	TEM/TER
Better understanding of the opportunities afforded by the railways as an alternative to road transport	T-H-T-R, TEM/TER
Better understanding of the opportunities afforded by water transport as an alternative to road transport	T-H-T-R
Competitiveness of the zone as a location for economic and other types of activities	T-H-T-R, TEM/TER, C 9
Identification of innovative effects of regional (subnational) development (e.g. spread of innovation, localisation of technological centres, road-access ports, telematic centres for traffic monitoring)	TEM/TER
Better understanding of the role of the zone within a European context	T-H-T-R, TEM/TER
Increasing interaction between planning levels (state-county-municipality), sectoral administration and states,	T-H-T-R, C 9
Growing understanding of strategic planning as learning process	T-H-T-R
Developing the benefits arising from transborder cooperation	TEM/TER-explicitly, C 9 and T-H-T-R indirectly
Staff-training and transfer of know-how	T-H-T-R, C 9
Technical objectives concerning infrastructure development	C 9, Blue Way
Synergy through planning	T-H-T-R, C 9

In these pilot projects the focus is on:

- environmentally sound economic development that also respects quality of life, cultural heritage, and other human needs,
- balance between road, water and railway transport,
- complexity of spatial development resulting from the interplay of different levels of authority (local, subnational, national, and international) with the participation of the business sector, non-governmental organisations, and other important players,
- growing spatial interdependence of European regions, which can no longer evolve separately; this means a significant change of emphasis from technical aspects to those addressing a social and economic impacts.

There is a desire to encourage similar work along the following corridors:

- Riga–Warsaw (southern part of via Baltica and transport corridor no. 1),
- St. Petersburg southwards up to corridor no. 1 with possible extension to Hamburg (via Hanseatica),
- Southern Finland–St. Petersburg–Moscow (northern part of corridor no. 9),
- Szczecin southwards (A 3) and the ferry lines to Ystad, Malmoe and Copenhagen,
- Berlin–Moscow from Orsha eastwards and from Brest westwards.

There is also a need to elaborate within a VASAB network the impact assessments of regional transport corridors complementary to the Trans-European Network (TEN).

Maritime Transport Cooperation

There is in addition a growing awareness amongst VASAB 2010 members that one of the key issues for future spatial development in the BSR is maritime transport and *port hinterland links, with a focus on intermodal transport centres.* Ten countries with considerable economic potential (in European terms) are located around the Baltic. Ferry traffic across the Baltic is the busiest in the world. Nonetheless, development of these links is currently being hampered by an inadequately evolved land-based infrastructure, particularly in the southern and eastern parts of the BSR. Combination of shipping with road, railway, and port development is an important aim of spatial planning in the Baltic region. Furthermore, the anticipated risk of lop-sided investment calls for prudent coordination and concerted action by Baltic spatial planners when it comes to developing shipping in the BSR.

There is also another challenge facing pan-Baltic spatial planning measures where maritime transport is concerned: How can environmentally friendly modes of conveyance be nurtured in the Baltic Region and how can efficient maritime transport with links to the potential sites of intermodal 'hubs' be developed if at the same time competition amongst urban centres increases. Cooperation on shipping within the VASAB 2010 framework has the aim of implementing the vision of sustainable development in the region as a whole, and there are moves afoot to gain EU participation (i.e. the European Spatial Development Perspective and G24 Group activities relating to the Baltic ports).

The detailed terms of reference of VASAB 2010 work in this field are now under preparation – with Sweden being appointed as the chair country.

Baltic Marinas

The Baltic Marinas concept elaborated within the Polish-German tourism development project is argued to be of great interest for all countries along the southern and eastern Baltic coasts and is therefore expected to encompass not only Poland and Mecklenburg-Vorpommern but also Kaliningrad, Lithuania, Latvia, Estonia and St. Petersburg. Yachting is clearly a huge potential attraction, growing in stature with increasing social affluence. The role of VASAB 2010 is to promote moves towards an international agreement between cities and towns hosting marinas in order to stimulate this form of tourism.

Spatial Planning, Natural Landscape and Cultural Heritage

One of the main assets of the BSR are its natural resources. As the coastal zone is particularly vulnerable, VASAB 2010 proposes the implementation of several projects under the umbrella of *the Common recommendations for spatial planning of the coastal zone in the Baltic region.*

In the northern and eastern parts of the BSR there are subregions where unspoilt forests, lakes and other biotopes are an important endogenous resource which can be used to stimulate the development of these areas. Sustainable tourism (eco-tourism) is one of the most frequently chosen alternatives of development in this part of the BSR.

The challenge for sustainable development of specific types of areas (patches) concerns how to maintain the existing high standard of the environment and biotopes whilst using these assets as the main leverage for development. To elaborate this question there is a need for an ongoing working dialogue between spatial planners and conservationists in which the needs of the local and subnational population will be taken into account. Accordingly, the VASAB 2010 objective in this field is to sustain dialogue and enhance mutual understanding between planners and environmentalists, so that a joint concept of development and conservation of areas of outstanding natural beauty or with a notable cultural heritage within the BSR can be worked out. The first step is to inventorise, classify, and, using a number of variables, evaluate natural assets and the cultural heritage.

Since developing eco-tourism is a priority for many Baltic Sea countries[4], VASAB 2010 intends to facilitate the creation of *a network of areas of natural beauty and cultural significance* which would form a Baltic Belt of Green Corridors – an attractive tourist complex benefiting tourists and residents alike. The network of Baltic Sea green areas is expected to create job opportunities and improve the accessibility of remote or peripheral areas while not being unduly onerous on the environment.

Cross-border Cooperation with a Spatial Planning Component

Two projects addressing cross-border cooperation with a spatial planning component have been conducted under the VASAB 2010 umbrella so far: *a spatial development framework for the Polish–German border region*, and cross-border cooperation between Latvia, Belarus, and Lithuania.

The first project involves drawing up a spatial development framework concept for the Polish-German border region and has been conducted jointly by the Polish and German ministries responsible for spatial planning. The framework concept was formally approved by the relevant ministries in both countries and by both national governments after intensive consultations with regional and other sectoral institutions. The spatial development framework facilitates cross-sector coordination of development measures, serves as guidance for transport infrastructure planning and private investors, and helps in promoting bilateral cooperation between private and public entities. Currently the project is at the implementation stage. The European Commission regards the German-Polish cooperation scheme as one of the most resounding successes in the transformation of spatial planning in Europe.

The *trilateral initiative between Latvia, Lithuania, and Belarus* is of particular importance for the coherence of the BSR since this is a region where national borders were re-instated in the early 90s. The main advantages of this project are the transfer of know-how in the field of advanced spatial planning, the activation of local authorities in peripheral Belarussian, Latvian, and Lithuanian regions, and the strengthening of cohesion in that part of the BSR experiencing ongoing problems of a structural nature.

Exchange of Knowledge

The seminar on National Strategic Spatial Planing Concepts in the BSR was held in September 1995 in Stockholm. The reason to hold such a workshop was growing demand for an exchange of knowledge in this field since most BSR countries have already drawn up their own national strategic planning documents

---

[4] See for example the *Denmark and European Spatial Planning Policy – National Planning Report* or Polish spatial development concept *Poland 2000+*.

and those who have yet to envisage doing so in the near future. During the seminar, the concepts already available (Sweden, Denmark, Finland, Poland) were presented and questions concerning the concepts chosen, to wit planning process and methodology, implementation, evaluation, and results achieved so far, were aired. This work will be continued in the form of training and technical assistance for BSR spatial planners at regional and local level. Regional links seem particularly important in this context.

For similar reasons, VASAB 2010 has initiated work on the BSR compendium of spatial planning legislation complementing EU efforts in this field.

Spatial Development Monitoring System

To monitor spatial development in the BSR and provide projections and analysis of ongoing spatial processes, it is necessary to have access to constantly updated information. This is already feasible, though access is limited and different standards and formats make transnational comparison difficult. Thus concerted action is needed to network existing databases and design them to meet the needs of spatial and subnational planners with the aim of capturing, standardising, and supplying all information of potential use to the VASAB 2010 programme and subnational planners throughout the BSR. VASAB 2010 also plans to facilitate professional contacts between spatial planners in the BSR and serve as a source of important information for them. Tasks should also include informing the public about major issues relating to spatial development in the BSR so as to increase the effectiveness of the decision making process in this field. Distribution of information via the Internet looks a very promising option here. This allows for unimpeded public participation in developmental and planning work.

Funding

In spite of undoubted successes, the shortage of funding for multilateral planning efforts has proved to be the main weakness at the implementation stage of VASAB 2010 work so far. This is mainly the result of:

- the weak position of spatial planning and other long-term and "soft" measures of development in transition countries preoccupied by short and medium-term problems at the macro-economic level,
- the conditions attached to existing international sources of funding, in particular Structural Funds, TACIS and PHARE, which are not designed for multilateral and network projects and cooperation.

In principle, the PHARE Cross-border Programme, PHARE Multi-Country Programmes and ECOS-OVERTURE could be used to fund spatial planning projects, but in practice there seem to be several factors militating against this. There is notably no solution for the concerted funding of projects executed jointly by

PHARE, TACIS and EU countries. The most frequent solution has thus been to use national finances (usually from Scandinavian countries and German national budgets) to execute such types of project.

INTERREG II C

Thus from the very beginning, and in tandem with work on the plan of action, there have been joint endeavours to secure reliable funding for VASAB 2010 projects, notably from international sources. These sources were targeted on the grounds that joint VASAB 2010 action has generated important benefits for a wider area than the Baltic region alone.

As a result of concerted action by the VASAB 2010 countries and the European Commission, a new Community Initiative named INTERREG II C was drawn up in 1996 to support transnational spatial planning throughout the EU. The transition countries were invited to participate in INTERREG II C work from the very beginning to allow them the opportunity to help shape the contents of the initiative.

VASAB 2010 is deeply involved in the preparation of the BSR INTERREG II C Operational Programme. The VASAB secretariat is represented in the taskforce preparing the document, whilst CSD/BSR members are involved in the process of harmonising it.

Thanks to this European Commission initiative, incentives will be created to encourage local, regional and even central bodies of the various countries to conjointly prepare and implement spatial projects compatible with the VASAB 2010 strategy adopted in 1994. The budget of INTERREG II C for the Baltic region is set to exceed ECU 40 million, half of which would come from EU structural funds and the other half from public and private sources at local, regional and national level. The sum is to be disposed of before the end of 1999. Under EU regulations, INTERREG II C monies can only be expended on the territory of the Union. Nevertheless, PHARE countries involved in multilateral projects to be financed within the framework of INTERREG II A and II C can also expect additional EU support. The Baltic PHARE CBC is to be increased by ECU 14 million for the years 1996–97, partly to offset INTERREG II C funds received by PHARE countries who are also participants in INTERREG II C projects. Preparations are also under way to create a TACIS CBC fund amounting to some ECU 30 million and devoted to backing transborder cooperation. That fund should be accepting the first applications before the end of 1997.

The Rostock Conference on Transnational Spatial Planning in the BSR

VASAB 2010 intends to facilitate the introduction of INTERREG II C into the BSR. Bearing in mind that, hitherto, Baltic Sea cooperation has focused predominantly on bilateral relations, making it more difficult to meet INTERREG II C provisions stipulating at least trilateral cooperation, the VASAB 2010 ministerial conference in 1996 decided to take advantage of the invitation extended by Germany to organise a conference in Rostock on 12 and 13 June 1997, the underlying theme of which could be briefly summarised as the "market of projects", i.e. a debate on concrete projects complying with the general proposals of the "Stockholm Declaration". During the conference, which was a kind of meeting between regional, local and national authorities and the representatives of VASAB, INTERREG II C and the European Commission, fully-fledged or planned spatial development projects were presented, while local and regional bodies had the opportunity to plug in to one or other of these projects or else initiate new multilateral projects with foreign partners in accordance with VASAB 2010 ideas and INTERREG II C requirements.

Harmonisation

Successful implementation of any regional development strategy can hardly be imagined without close cooperation between the various players and stakeholders. Thus from the very beginning VASAB 2010 has worked together with a number of regional interest groups and authorities at differing levels. the most important of these being the Commission of the European Union, the Council of the Baltic Sea States, the Conference of the Baltic Sea Ministers of Transport, HELCOM, the Baltic University, the Subregional Conference etc. In 1996 the Baltic Sea ministers for the environment drew up their own programme aimed at sustainable development for the Baltic region – "Agenda 21 for the BSR" – which seems to be the first attempt to create Agenda 21 at regional level.

Agenda 21 for the Baltic Sea Region

The main idea behind Agenda 21 for the BSR is to provide the region with some kind of horizontal coordination of sectoral policies with the aim of improving the sustainability of BSR development. The work is organised around sectoral studies in the following fields: energy, industry, transport, tourism, agriculture, forestry and fisheries. These analyses have the objective of integrating sectoral policy and environmental concerns and will be conducted jointly by sectoral and environmental experts. Integrated approaches such as VASAB 2010 or HELCOM will also be taken into account. The process is being overseen by the Group of Senior Officials (SOG) comprising representatives of the environment ministries for the Baltic Sea countries and non-governmental and intergovernmental organisations (including VASAB 2010). The SOG intends to present an Agenda 21 concept for the BSR in early 1988.

Since the Agenda 21 for the BSR has aims similar to those of VASAB 2010 and complements VASAB 2010 work, it was always clear that both organisations should work hand in hand. VASAB 2010 has participated in the Agenda 21 process from the very beginning, supporting the idea in the declaration of the ministerial conference and ratifying SOG membership. The collaboration between VASAB 2010 and Agenda 21 is an excellent example of precisely how important networks involving the various influential players are becoming for the development of European regions like the Baltic.

Conclusions for the Black Sea Region

The Black Sea Region has not yet commenced work on a long-term strategic concept of sustainable development. This can be also seen as an advantage since it enables the region to capitalise on existing VASAB 2010 experience. The following conclusions would appear to be of relevance to the Black Sea Region here:

1. Regional cooperation in the field of spatial planning would appear to bring important long-term benefits for all countries in the region, including countries in transition. An agreed common vision of spatial development can improve the image of the region, attract extra investment from the private sector, and identify any "spatial bottlenecks" in regional development.
2. Amidst growing competition between European regions, a common agreed vision and its concerted implementation at regional level improve a region's chances of accessing the limited pool of resources earmarked for regional and spatial development.
3. Regular working contacts between key persons in the spatial planning domain also produce immediate benefits. For example they engender mutual understanding and better coordination of cross-border work in the field of spatial planning. In such an environment it is easier to come to decisions on difficult and controversial issues of a multilateral character. It also facilitates the transfer of know-how and consolidates the identity of the region.
4. Unfortunately, added value from spatial planning is not markedly in evidence for all regional stakeholders, especially for those from transition countries preoccupied with ensuring macro-economic stability. Therefore it is important to start a VASAB-like cooperative scheme by informing stakeholders of the short and longer-term benefits of such initiatives.
5. A strong political will and political engineering are necessary to start such a form of cooperation. Without a "political" umbrella such cooperation can hardly hope to be successful.
6. It is difficult to conceive of multilateral cooperation without common consensus concerning the goals and nature of development in the region. However,

such consensus is even more important at the implementation phase when resources have to be allocated to different ventures. The VASAB 2010 experience shows that such consensus and mutual understanding are possible after a period of joint endeavour notwithstanding differences among the cooperating countries with regard to their level of development and other characteristics.

7. Spatial planning is a learning and interactive process where different players learn how to cooperate with each other whilst respecting different interests and backgrounds. The key notion is interaction. Thus, to be effective multilateral spatial planning should be part of a broader network of regional cooperation. The same is true for both the national and the subnational levels. The VASAB 2010 experience is that only close collaboration between subnational entities, local authorities and sectoral ministers can ensure successful implementation of pilot projects and joint action. The VASAB 2010 national committees have been very helpful in this respect.

References

Zaucha J (1997) "From Vision to Action" drafted by Jacek Zaucha based on discussion with the Committee on Spatial Development of the Baltic Sea Region in "Vision and Strategies around the Baltic Sea 2010 – From Vision to Action", June 1997

Sustainable Development and Coastal Zone Management around the Baltic Sea

Common Recommendations for Spatial Planning of the Coastal Zone in the Baltic Sea Region

Helle Fischer
Ministry of Environment and Energy
Spatial Planning Department
Copenhagen, Denmark

Introduction

Ministers responsible for spatial planning and development from all the countries around the Baltic Sea met at the Fourth Conference of Ministers on 22 October 1996 in Stockholm. One of the main topics at the conference was the discussion and adoption of a proposal concerning common guidelines for spatial planning in the coastal zone entitled *Common Recommendations for Spatial Planning of the Coastal Zone in the Baltic Sea Region.*

All the ministers approved the proposal. Each individual country is now to implement the coastal zone recommendations in national policies and legislation.

This article sets out the background to the initiative as well as the purpose of common spatial planning recommendations for the coastal zone in the Baltic Sea Region. This is followed by a more detailed description of the individual recommendations. Finally the recommendations are put into perspective and their importance for future physical and spatial development in the coastal zone of the Baltic Sea Region is assessed.

Background

VASAB 2010

The move to draw up common recommendations for spatial planning in the coastal zone is part of a major strategy and action programme covering the physical and functional development of the entire Baltic Sea Region. In 1992 the planning ministers of all the countries around the Baltic Sea initiated the process of formulating a spatial development perspective document. The working group comprised representatives of the planning authorities in Belarus, Denmark, Estonia, Finland, Germany (federal level plus Mecklenburg-Vorpommern and Schleswig-Holstein), Latvia, Lithuania, Norway, Poland, Russia (Kaliningrad, St.

Petersburg) and Sweden. A report entitled *Vision and Strategies around the Baltic Sea 2010 (VASAB 2010)* was presented and adopted, together with proposals for 21 specific items of action, at the Third Conference of Ministers at Tallinn in December 1994.

The report sets forth a framework for spatial development in the Baltic Sea Region (BSR), the purpose of which is to ensure that "the Baltic Sea Region develops a strong identity based on sustainability, enhanced spatial cohesion and cross-border cooperation. Vision and strategies shall assist to harmonise the spatial aspects of the various policies, programmes and initiatives in the region".

The vision portrays future development in spatial terms, dividing it into three basic units of spatial structure:

the *pearls*	cities and urban networks
the *strings*	mobility networks and energy supply (ports, railways, roads, airports, telecommunications, pipelines, electricity)
the *patches*	specific types of areas (coastal zone, islands, border areas, agricultural areas, valuable cultural landscapes, forestry areas, wetland biotopes, protected areas)

and the overarching planning process:

the *system*	spatial planning (institutional and procedural approaches and problems of spatial planning in the various countries).

Strategies and initiatives designed to put this vision to effect are formulated for each unit of spatial structure and planning process. All 21 items of action proposed were approved by ministers. Some have a regional dimension and will only involve a few countries, while others cover the entire Baltic Sea Region.

A Committee for Spatial Development in the Baltic Sea Region composed of senior officials (henceforth referred to as the VASAB Committee) was established by the ministers to ensure implementation of the 21 items of action and the deepening of cooperation within the Baltic Sea Region[1].

One item in which all participating countries are involved is the project to "elaborate guidelines for spatial planning in the coastal zone", which is based on the vision that "the coastal zone be planned maintaining a careful balance between development and protection".

[1] A more detailed description of VASAB 2010 cooperation is provided in this publication in the article "VASAB 2010 – Transnational Cooperation in the Spatial Development of the Baltic Sea Region" by Jacek Zaucha, Deputy Secretary, VASAB Secretariat, Gdansk, Poland.

Why do We need Common Guidelines for Spatial Planning in the Coastal Zone?

The coastal areas in the Baltic Sea Region are characterised by significant regional differences in terms of land use and the pace of development. The scenario is constantly changing. Together with a number of other factors, the dramatic political changes in the Baltic Sea Region following the fall of communism, the independence of the three ex-Soviet Baltic States, and the EU membership of Sweden and Finland will all affect the use of landscapes. Increasing levels of economic activity will lead to the further earmarking of land for urbanisation, industrial installations, infrastructure and facilities for tourism in coastal areas.

In Denmark one third of the coastline consists of built-up areas or areas which have been set aside for development activities. Half of the coastline is visually affected by this. The same phenomenon is to be observed around all capital and major coast-line cities in the countries bordering the Baltic, e.g. Stockholm, Helsinki, St. Petersburg, Tallinn, Riga, Klaipeda, Gdansk-Gdynia, Rostock, Lübeck, Gothenburg and Oslo.

Along the Gulf of Bothnia and the entire eastern part of the Baltic, by contrast, there are enormous areas with few or no built-up areas. Some of these areas are used for agriculture or forestry, while others have been designated as various types of protected or set-aside areas. In the eastern part of the Baltic Sea Region in particular it is still possible to find 'virgin nature' where flora and fauna remain virtually unaffected by human activity.

The tourism industry is one of the fastest growing industries in the world. In many countries tourism is a major contributor to the national economy, both in terms of jobs and foreign exchange earnings.

Clearly, large sections of the Baltic coastline have an appeal that could be far more heavily exploited than is the case today. The nature and culture of the Baltic Sea Region are in many ways different from those of the rest of Europe. The region is characterised by white, sandy beaches alternating with rocky coasts, islands and archipelagos of unique beauty, by medieval towns and villages of great architectural value, and by small, picturesque fishing villages located in the coastal zone or very close to it.

A fundamental precondition for ensuring that these attractions become an active asset enhancing both the standard of living of the local population and future tourist-sector development is that both the cultural and natural environments be safeguarded by means of properly thought-through functional and area-related planning. The uses to which coastal areas in southern Europe have been put, where developments in the tourist industry have often been feverish and unbridled, serve as a useful lesson as to what can happen if planning instruments do not include measures to ensure territorial exploitation is adequately regulated.

The transition countries are facing major challenges as they address themselves to the future development of their coastal areas, a process that also opens up numerous opportunities. The challenges of the future revolve around the need to take appropriate measures to ensure the long-term socio-economic development of coastal areas, promote environmental sustainability, and protect the natural environment and cultural heritage.

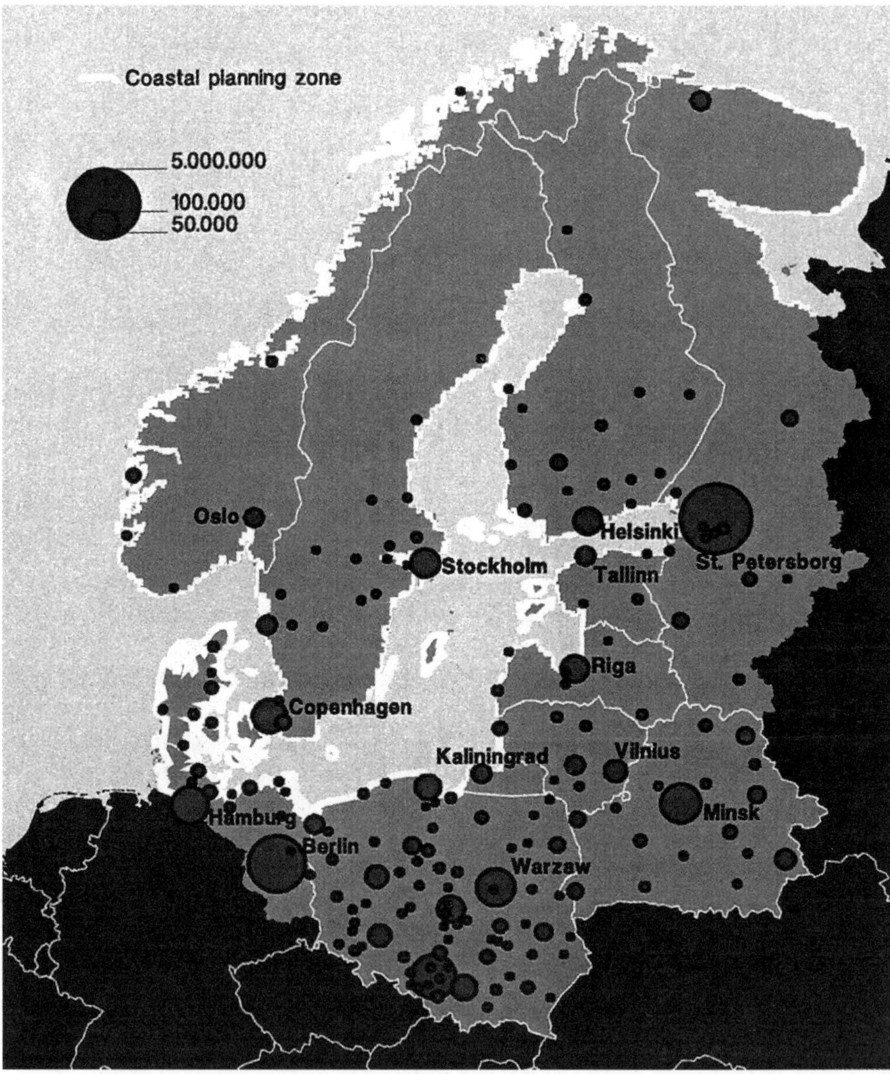

Figure 1. The Baltic Sea Region coastal planning zone (3 kilometres) and urban centres (> 50 000 inhabitants)
Source: From the report "Vision and Strategies around the Baltic Sea 2010".

Existing Spatial Planning Legislation Applicable to Activities in Coastal Areas in the Baltic Sea Region

Most BSR countries have legal provisions regulating activities that may have an impact on the natural environment in the coastal strip outside urban areas. The width of this protection zone varies from one hundred to three hundred metres landwards from the shoreline. In all countries this zone is a protected area in which clear and often very stringent restrictions on construction and development or, indeed, complete bans on construction apply. In several countries this zone or borderline area is designated a "nature protection area". Sweden and Latvia also have a protected off-shore strip.

There are no clear guidelines however relating to the area behind this protection zone. This means that the construction of new housing, hotels and technical installations behind the protection zone is not governed by any specific rules. This may have a crucial impact both on the visual effect of the coast and on public access to coast, beach, or sea.

The intention behind the new coastal guidelines is to define a zone of a different nature than the existing protection zone of 100–300 metres, allowing instead specific conditions and development potential in individual coastal areas to be taken into account. The idea, in short, is to formulate guidelines which regulate while at the same time accommodating divergent interests within the coastal zone. Consequently, the intention has not been to define a new protection zone but rather a new type of planning zone[2].

The Danish Experience

In Denmark, development in coastal areas has been monitored for many years. Over the past 40–50 years an increasing proportion of coastal areas has been actively made use of for various purposes. Towns and urban conglomerations have spread along the coast, areas with summer houses and holiday centres have sprung up in locations close to the coast, and technical installations such as power plants and waste dumps have grown in size and assumed a higher profile in the coastal zone than was previously the case.

The threat imposed by developments in the coastal zone moved the Danish parliament to decide to include a number of special provisions in the Danish Planning Act relating to activities within a three-kilometre zone outside existing urban zones as well as a number of special regulations for those sections of urban zones situated close to the coast. At the same time the Danish parliament decided to extend the coastal protection zone from 100 metres to 300 metres on the grounds that the 100-metre strip was too narrow.

---

[2] A definition of the term planning zone is given in the section: Common recommendations for spatial planning of the coastal zone, p 187.

The purpose of the new provisions of the Planning Act is to limit the construction of houses and other buildings in the coastal zone, especially along open stretches, and to ensure that buildings which are to be sited close to the coast are located in such a way as to ensure that their visual impact on the coastal landscape is no greater than need be[3].

Given the Danish experience, Denmark offered to act as the main promoter for the initiation and formulation of common guidelines for spatial planning in the coastal zone of the Baltic Sea Region, in close cooperation and consultation with all participating countries. Material relating to coastal planning that had already been enshrined in Danish legislation has thus been taken into account in the formulation of the recommendations for the entire Baltic Sea Region.

Coastal Recommendations

The issue of development and management in the coastal zone has also been on the agenda elsewhere in the international arena in recent years. This can, for example, be seen from the number of international programmes and activities devoted to coastal zones.

Integrated Coastal Zone Management is an integral, cross-sectoral method embracing all sectors that have an impact on a given coastal zone. The OECD has applied this method to the issue for several years and is now preparing a Review of progress towards integrated coastal zone management in selected OECD countries. In February 1996 the EU Commission launched a Demonstration Programme on Integrated Management of Coastal Zones.

The VASAB Committee has followed the international debate closely. Analyses and regulations have been used as an inspiration for, or else have been directly input into, coastal recommendations.

The most important item of input is the coastal zone definition contained in HELCOM[4] Recommendation 15/1, Protection of the coastal strip, adopted in March 1994.

HELCOM Recommendation 15/1 expresses a "deep concern about the increasing pressure for commercial exploitation of the remaining natural coast, particularly in countries in economic transition and countries with no legislation for the protection of the coast, and about the consequences this might have for the coastal environment and the biological diversity of the whole Baltic area".

This recommendation makes it clear that any improvement of the marine environment in the Baltic Sea needs to be seen in the context of environmental improvement of land-based activities. In this connection a three-kilometre planning

---

[3] Spatial Planning Department of the Danish Ministry for the Environment and Energy published in June 1997 "Spatial Planning in the Coastal Zone in Denmark".

[4] HELCOM (The Helsinki Commission, working with the "Protection of the Marine Environment in the Baltic Sea Area").

zone is defined within which "major building development and other major permanent changes in nature and landscape be preceded by an appropriate land use plan, including impact assessment, approved at least on regional level".

In addition to HELCOM Recommendation 15/1, relevant parts of HELCOM Recommendation 15/5 of 10 March 1994 concerning a "System of coastal and marine Baltic Sea protected areas (BSPA)" and HELCOM Recommendation 16/3 of 15 March 1995 concerning the "Preservation of natural coastal dynamics" have been incorporated into the VASAB recommendations.

The overall goal for VASAB cooperation in the Baltic Sea Region is to create a coherent, well-functioning region based on development, environmental sustainability, freedom and solidarity. It comes as little surprise, therefore, that the VASAB Committee has taken on board the statements and initiatives contained in the United Nations final document from the Rio de Janeiro conference in 1992 – 'The Rio Declaration' – and in Agenda 21, notably Chapter 17 of the latter. Inspiration and input were also forthcoming from the World Coast 1993 Conference, which expresses the urgent need for coastal states to develop and strengthen their capabilities for integrated coastal zone management, and from the UN's Washington Declaration on the Protection of the Marine Environment from Land-based Activities, adopted in November 1995.

Common Recommendations for Spatial Planning of the Coastal Zone in the Baltic Sea Region[5]

Sustainable development in the coastal zone is a fundamental goal. The coastal zone must be planned in such a way as to ensure a careful balance between development and protection. The common recommendations for spatial planning in the coastal zone must be sufficiently flexible to enable the most expedient localisation of new activities in the coastal zone to be recommended, but they must at the same time be sufficiently restrictive to shield the greatest possible coastal area along the Baltic Sea and keep it free from new urban development, hence minimising degradation of the existing cultural and natural heritage.

Structure and Content

The purpose of the recommendations has been to formulate a set of rules – a kind of toolbox – which can be used and incorporated directly into national and regional/local planning for coastal zones.

The common recommendations – or guidelines – are divided into four parts: preamble, objectives, recommendations and planning procedures.

[5] The full text of the Common Recommendations for Spatial Development of the Coastal Zone in the Baltic Sea Region is appended to this article.

The preamble describes the need for such guidelines and explains why it is important to monitor and control future physical and spatial development in coastal zones.

The objectives express the desired directions and visions for future development:
to protect yet develop
to strengthen economic and social development in a sustainable manner
to guarantee biodiversity
to protect and conserve the natural and cultural heritage.

The recommendations contain a list of regulations and provide *guidance* for the process of spatial planning and management in the coastal zone. This is the most important part, since it specifies the planning tools. Selected issues are explained in greater detail below.

Definition of the Coastal Zone

All countries participating in VASAB are also parties to HELCOM. From the beginning, the VASAB Committee was therefore bound by definitions already adopted in HELCOM Recommendation 15/1. The question was then whether a three-kilometre planning zone was wide enough from a planning point of view. Proposals for a five-kilometre zone were tabled. Also the size of the small totally protected strip was discussed. The VASAB Committee concluded that a three-kilometre planning zone would be able to meet the objectives and therefore decided to adopt the HELCOM definition as it stood.

The planning zone around the Baltic Sea is now defined in accordance with HELCOM Recommendation 15/1:
a planning zone of at least 3 kilometres landwards
a universally protected strip outside existing urban areas and existing settlements of at least 100–300 metres landwards and seawards from the mean water-line.

In addition to planning considerations, the VASAB Committee believes it will only be possible to achieve future objectives through cooperation and coordination. Moreover, the application of common objectives and definitions by various authorities and institutions wherever such common objectives and definitions are appropriate could give rise to a synergy effect. The development of the coastal zone will benefit from a more cross-sectoral approach to the solution of spatial and environmental problems.

Comprehensive Plans

In order to ensure functional and sustainable development of the coastal zone, activities in the coastal zone should be based on overall national and regional plans determining the overall framework of land use. These plans must take overall long-term interests within society into account and must ensure functional cohesion between future development in the coastal zone and in the country concerned in general. More detailed plans should be drafted at appropriate levels in order to ensure the participation of local authorities and the general public.

Functional Justification

One of the most important principles underpinning the recommendations is that of keeping the coastal zone as free as possible of construction and infrastructure ventures that may be expected to jeopardise natural and cultural resources and recreational interests. Planning for new activities involving urban development, construction, infrastructure, vacation centres and leisure facilities outside existing urban settlements in the coastal zone should therefore be based on either a planning-related or a functional justification for a coastal location.

A *planning-related justification* could obtain for the expansion of existing cities or towns in cases where the location of new urban areas outside the coastal zone is not feasible geographically or would conflict with important planning considerations. Extensions to infrastructure and public service functions in which investments have already been made, and where a stop would result in significant economic losses, are a further example. Finally, mention needs to be made of urban development on smaller islands and in archipelagos, notably in cases where the need to ensure the development and survival of local communities outweighs the wish to protect coastal areas.

A *functional justification* could obtain in the case of installations requiring a location close to the coast for various reasons: marine traffic or fishing activities, access to cooling water or wind (e.g. windmills), or waste deposit sites (in order to protect groundwater resources further inland from potential contamination).

Keeping Open Stretches of Coast Free of Construction

A comparison of topographical maps of the Baltic Sea Region issued forty years ago and similar recent maps show the extent of urban sprawl along the coasts in the intervening period. The prime objective now is to preserve open stretches of coast that survive. It is therefore imperative that urban development take place predominantly inside, close to, or behind existing settlements so as to avoid urban sprawl along the coastline.

Green wedges or strips inside cities or between cities located close to one another should be maintained or even newly established in order to ensure that new urban development will not create a visual barrier along the coast.

Impact Assessment

Large construction ventures, infrastructure projects, land reclamation and other major permanent alterations to the coastal zone need to be preceded by either an appropriate regional or local land-use plan or a special regional or local impact assessment covering both visual and environmental aspects.

Windmills

One of the objectives of VASAB is to further the use of renewable and environmentally friendly sources of energy. Energy produced by windmills is a renewable form of energy which in most countries would be seen as a functional justification for a location close to the coast. When adequate areas with sufficient wind resources have been designated, it should be ensured that the location does not conflict with existing plans and is not detrimental to sensitive nature and bird protection areas. A concentration of windmills in groups or windmill parks may also ensure a more efficient use of specified areas, thus obviating the need for windmills in coastal areas with a more valuable natural environment or cultural landscape.

Sewage and Waste Deposit

The designation of areas for the discharge of sewage and for waste dumping is a general problem. The choice of location depends on a number of factors including geography, groundwater and the environmental impact on the drinking water. There are significant variations from locality to locality.

The discharge of untreated sewage should be avoided in the coastal zone, and waste deposit sites should only be allowed following approved land-use plans and once appropriate regulations such as regional and environmental impact assessment requirements have been met.

Natural Values

There are huge areas in the Baltic Sea Region which are rich in natural values. A small proportion of them have already been lent the status of nature conservation areas, but the majority are now being used for farming or forestry purposes. Future agricultural reforms and changes in ownership and production patterns, particularly in transition countries, will undoubtedly have an impact on the layout of landscapes. These areas should therefore be incorporated into a 'green' infrastructure as part of a policy of sustainable development of human settlements.

Vacation Centres and Leisure Facilities

Vacation centres and leisure facilities include holiday hotels, camping sites, youth hostels, marinas, golf courts and leisure parks. The location of such facilities must be based on an overall national or regional tourism policy and must be in keeping with local land-use plans. Centres must fit into the landscape in such a way as to ensure that the recreational values and the natural beauty of the coastal zone is preserved and that continued public access to the coast and coastline is ensured.

Cultural Heritage

In many of the towns and rural areas in the Baltic Sea Region, historical developments stretching back to the Middle Ages can be made out. It is vital that this cultural heritage be preserved. Instead of planning new building areas, historical settlements need to be refurbished and restored.

Figure 2. Urban development needs to take place inside, close to or behind existing settlements to avoid 'urban sprawl' along the coastline.

Figures 3 und 4.
Coastal areas close to urban centres or inside cities have an important recreational value for the citizen. Therefore, green wedges or green strips should be maintained or, indeed, new ones established to ensure continuing public access to the coast.
Source: Vejledning om planlægning i kystområderne, Ministry of Environment and Energy, 1995.

Planning Procedures

The coastal recommendations conclude with a number of practical guidelines concerning future development, including implementation and monitoring. By adopting these coastal recommendations the ministers have advocated that the objectives, the recommendations and the planning procedures relating to spatial planning of the coastal zone should be included in national legislation and policies. The implementation of recommendations is regarded as a national, regional or local issue subject to the principle of subsidiarity and the division of responsibilities in individual countries.

Assessments

Protection and Development

The formulation of the recommendations may convey the impression that more emphasis is placed on protection than on development. This is not the case. On the contrary. The recommendations are not to be used to stop desirable development ventures in the coastal zone but rather to point up and determine what should *not* be done, the purpose being to ensure *positive and sustainable development.*

The recommendations are intended to be of use both for countries such as Denmark, which requires very restrictive management of the coastal zone, and for countries such as the transition countries, which are in the midst of a restructuring and development phase.

Conflicts Relating to Land Use

The recommendations may also be used to prevent disputes. Conflicts of interest may easily arise in attractive coastal areas where some parties may want to construct new housing, business facilities or tourist facilities, while others may want to preserve the special natural values in the area concerned.

Coastal Planning in Urban Zones

The majority of the population in the Baltic Sea Region lives in cities or small towns[6]. A large proportion of these towns and cities are located in the coastal zone, with the exception of towns and cities in Poland and Germany. Owing to the still high share of those employed in the primary sector in the transition countries (> 20 %), a considerable increase in urbanisation in these countries is to be expected.

---

[6] The urbanisation rate varies between 58 % in Estonia, 62 % in Poland and Finland, 86 % in Sweden and 91 % in St. Petersburg Oblast (figures from 1992/93, the VASAB 2010 report).

The high density of population in coastal regions and the expected growth in urbanisation in the transition countries will mean increased pressure on towns and cities located close to the coast.

However, the recommendations do not address themselves to towns and cities. At present there is no common interest in the Baltic Sea Countries in formulating guidelines for spatial planning in the coastal areas of urban zones.

Political Opinions

As mentioned in the introduction, the *Common Recommendations for Spatial Planning in the Coastal Zone in the Baltic Sea Region* were tabled at the fourth Conference of Ministers in the Baltic Sea Region held at Stockholm on 22 October 1996. There was broad political support for the recommendations from all participating countries.

Perspectives

It is very likely to take some years before the recommendations have been adapted to and incorporated into national and/or regional legislation on spatial planning. However, this should not prevent countries from heeding the recommendations immediately. The ministers thus asked the VASAB Committee to initiate pilot projects covering diverse uses of coastal areas in order to gain practical experience in this field.

The VASAB Committee's financial resources are fairly modest[7]. The number of projects which the Committee will be in a position to initiate is hence limited, as is the scope of projects that can be initiated.

The European Commission's new INTERREG II C[8] initiative, which is a programme of "Transnational Cooperation in Spatial Development", is an important instrument in this respect that provides fresh opportunities.

At the Stockholm Conference in October ministers emphasised in their Declaration that "the European Commission should establish the necessary precondi-

---

[7] Annual budget of about DKK 2–2.5 million, which is also to cover costs relating to the VASAB Secretariat in Gdansk.

[8] Sweden, Finland, Germany and Denmark have formulated a common Operational programme for the Baltic Sea Region, which was submitted to the European Commission, DG XVI, in February 1997. The Programme was formulated in close collaboration with Norway and the VASAB committee (the VASAB Secretariat participates in all INTERREG II C meetings as an observer). It is expected that negotiations concerning the Programme for the Baltic Sea Region will be concluded during the summer, so that the application deadline for the first projects can be set in early autumn 1997.

tions to ensure that joint development initiatives and projects in the Baltic Sea Region can be financed out of funds provided within the framework of both the INTERREG II C Community Initiative and the PHARE and TACIS support programmes for countries in transition. Only by integrating these two support mechanisms of the European Commission will it be possible to carry out transnational development projects initiated by the Member States of the EU and other countries along the Baltic Sea and designed to promote the spatial cohesion of this region".

However, it is still uncertain whether the European Commission will assent to INTERREG II C being coupled with PHARE and/or TACIS. It would then be up to regional and local authorities to apply for and initiate transnational coastal planning projects.

It will take a few years before there is any basis for an assessment and evaluation of the importance of the recommendations on sustainable development in the coastal zone. In the meantime the VASAB committee will, as requested by the ministers, initiate systematic monitoring of development in the coastal zone and organise exchanges of information between the countries concerned.

Common Recommendations for Spatial Planning of the Coastal Zone in the Baltic Sea Region

Fourth Conference of Ministers responsible
for Spatial Planning and Development
in the Baltic Sea Region
Stockholm

Adopted 22 October 1996,

THE MINISTERS
responsible for Spatial Planning and Development from the participating countries in the "Committee for Spatial Development in the Baltic Sea Region",

BEARING IN MIND the Resolution from the Third Conference of Ministers for Spatial Planning and Development of the Baltic Sea Region, Tallinn, 7–8 December 1994 in which the Ministers welcomed the report entitled "Vision and Strategies around the Baltic Sea 2010" and approved the proposed items of common action,

BEARING IN MIND IN PARTICULAR item of common action 4.3 in "Vision and Strategies around the Baltic Sea 2010" concerning elaboration of "guidelines for spatial planning in the coastal zone",

BEING CONSCIOUS that the coastal zone in the Baltic Sea Region has great potential in terms of cities, harbours, industries, infrastructure, valuable resources, natural and cultural heritage which can be used for economic and social development, that it is at the same time a unique yet vulnerable area which needs protection and that the natural processes in the coastal zone are dynamic and subject to transboundary human activities,

BEING ALSO CONSCIOUS that the expected expansion of cities, tourism areas, transport and technical installations especially in the southern and eastern part of the Baltic Sea coastal zone needs to be steered by means of regionally balanced sustainable development,

DRAWING ATTENTION TO the fact that spatial planning has a crucial role to play in coastal zone management as a means of supporting economic and social development and cooperation along the coast, to contributing to the coordination and integration within a regional context of sectoral measures of national and local interest, and establishing a multisectoral framework for land use to balance the needs of environmental protection and development,

REGARDING coastal zone management as a broad social, economic and ecological approach with the aim of improving coordination and harmonisation of all planning and implementation activities that significantly impact on the quality of the environment, economic and social opportunities and the cultural heritage in the coastal zone,

RECOGNIZING the need to elaborate common guidelines for spatial planning and management in the coastal zone in order to both ensure and channel spatial development in this area,

BEING AWARE that the coastal zone must be planned maintaining a careful balance between development and protection, with the implication that regulations and legislation relating to the coastal zone must at one and the same time promote economic and social development and ensure the safeguarding of nature, of areas of biological diversity, of landscapes and the cultural heritage and of public access and free movement to and along the coast,

TAKING INTO ACCOUNT HELCOM Recommendation 15/1 concerning protection of the coastal strip adopted 8 March 1994, HELCOM Recommendation 15/5 on the system of coastal and marine Baltic Sea Protected Areas (BSPA) adopted 10 March 1994, HELCOM Recommendation 16/3 on the preservation of natural coastal dynamics adopted 15 March 1995, the UNCED resolutions passed at Rio de Janeiro in 1992, and notably Chapter 17 of Agenda 21, statements from "The World Coast 1993" conference, in which expression is given to the urgent need for coastal states to develop and strengthen their capabilities for integrated coastal zone management, and, finally, the Washington Declaration on the Protection of the Marine Environment from Land-based Activities, adopted 1 November 1995,

TAKING ALSO INTO ACCOUNT the existing provisions and legal acts concerning spatial planning and protection of the coastal zone in all the participating countries/regions,

APPRECIATING the measures already taken by several Baltic Sea countries to protect their coastal zones,

THE MINISTERS or their representatives,

RECOMMEND that the following objectives and planning procedures concerning spatial planning of the coastal zone be included in national legislation or policies:

I Objectives

a) actions in the coastal zone should promote sustainable development, ensure appropriate living conditions for residents, and secure a dynamic balance between the coastline, biological diversity, and the cultural heritage,
b) the economic potential of the coastal zone should be maintained and developed, i.e. the development of cities, settlements, economic activities and infrastructure in the coastal zone should be economically, socially and ecologically sustainable,
c) conservation of biological diversity, the cultural heritage and the sustainable use of natural resources on the landward and seaward sides of the shoreline are considered a prerequisite for sustainable economic growth and development of the coastal zone,
d) the coastal zone should be kept as free as possible of construction and infrastructure ventures that may be expected to jeopardise natural and cultural resources and recreational interests,
e) land use and the scale of buildings and changes in the physical environment in the coastal zone should take account of the cultural heritage and the natural and physical conditions of the landscape, especially with regard to archipelagos, islands, spits of land and sandbanks,
f) the cultural heritage and nature and landscape values in the coastal zone should be preserved through appropriate planning and management,
g) the impact of land reforms on landscape values in the coastal zone should be minimised,
h) the recreational values and the natural beauty of the coastal zone should be safeguarded and continuing public access to the coast and along the coastline ensured,
i) the elaboration of comprehensive plans for urban areas, environmentally friendly technologies, transport systems and energy resources should be encouraged,

II Recommendations

concerning channelling the process of spatial planning and management in the coastal zone:

a) the coastal zone around the Baltic Sea shall be defined in accordance with HELCOM Recommendation 15/1, which urges that a planning zone of at least 3 kilometres landwards and a universally protected strip outside existing urban areas and existing settlements of at least 100–300 meters landwards and seawards from the mean water-line be established,

b) comprehensive plans aimed at supporting sustainable development in the coastal zone should be drafted at the appropriate levels (national, regional, local) ensuring the participation of local authorities and the general public,

c) planning for new activities relating to urban development, construction, infrastructure, vacation centres and leisure facilities outside existing urban settlements in the coastal zone should be based on either a planning-related or a functional justification for a coastal location,

d) location of new buildings and constructions should preferably be effected inside, close to or behind existing settlements to avoid urban sprawl along the coastline and maintain the natural landscape,

e) new developments should not form a visual barrier along the coast, therefore green wedges or strips should be maintained or provided,

f) any large building venture, infrastructure project, land reclamation measure or other major permanent alteration to the coastal nature and landscape should be preceded by either an appropriate regional or local land-use plan or a special regional or local impact assessment covering both visual and environmental aspects,

g) windmills should be located outside areas important for migratory birds and in such a way that they minimise the impact on valuable landscapes,

h) comprehensive and/or local land use plans should be elaborated as a matter of particular urgency in sensitive and/or congested areas,

i) options for improving hinterland links to existing harbours and allowing future revitalisation of existing harbour areas should be kept open,

j) discharge of untreated sewage in the coastal zone should be avoided and location of waste deposit sites should only be allowed given approval of land-use plans and compliance with appropriate regulations such as regional and environmental impact assessments,

k) areas with varying natural values, whether they are worthy of preservation or not, should be incorporated into a green infrastructure in order to create a better concept for sustainable development of human settlements,

l) vacation centres and leisure facilities should be located in line with local land use plans and the coherent requirements of national or regional policies on tourism taking into account the preservation of landscapes, nature, cultural heritage and the carrying capacity of said landscapes,

m) coastal protection measures outside settlements should only be initiated where they are considered necessary for preserving the natural coastline,

n) environmentally friendly transport and energy systems should be encouraged and the problems caused by private motoring in sensitive coastal areas should be given due attention,

o) the cultural heritage and characteristics of the coastal settlements should be maintained and renewed, and re-establishment of historical settlements should be preferred to new building areas,

III Planning Procedures

a) the Ministers shall take appropriate measures to ensure implementation of the recommendations for planning in the coastal zone around the Baltic Sea,

b) the implementation of recommendations for planning in the coastal zone is regarded as a national, regional and local issue in accordance with the principle of subsidiarity and the division of responsibilities in a given country,

c) regional impact assessments, including environmental impact assessments, should be applied in accordance with national legislation, and information for the general public should be provided at a very early stage of the planning process for any large project having a significant impact on spatial development and the environment,

d) development in the coastal zone should be subject to systematic monitoring, and information should be exchanged among the countries concerned,

e) each country is responsible for monitoring its own territory and for the exchange of information concerning installations which might have a transboundary effect (The Espoo Convention),

f) special areas should be designated, when appropriate, with the objective of conserving nature and safeguarding marine and coastal biodiversity, after the example of the Baltic Sea Protected Areas (BSPA).

FURTHERMORE the ministers recommend that the Committee for Spatial Development in the Baltic Sea Region monitor the implementation and the outcome of recommendations through demonstration projects and report to the next Ministerial Conference.

Ecological Prospects for the Black Sea Region

Alexander S. Shilov
Russian Academy of Public Administration
under the President of the Russian Federation
Moscow, Russia

The Black Sea is on the verge of ecological catastrophe. The marine ecosystem is threatened with destruction. The sea and land close to the sea are polluted with oil products, heavy metals, organic fertilizers and radioactive materials. The biological resources of the sea are decreasing at a disastrous rate, and many kinds of fish and sea products are no longer fit for human consumption or are actually harmful.

The sea accumulates various negative consequences of human activity in its vast stores of water. The status of the Black Sea, as determined by the Consulting Committee for the Protection of the Seas, is acknowledged to be that of a "zone of ecological disaster". A worsening of the ecological situation would have serious implications not only for countries in the region but for the biosphere potential of the entire Mediterranean basin. The critical ecological situation that has emerged in the Black Sea ecosystem compels us to pay special attention to a number of mutations that have occurred in the process. Millions of people are immediately affected by the Black Sea, and as such the state of the natural environment in the region is a matter of vital importance for them.

When analysing the ecological prospects of the Black Sea region, it is first of all necessary to stress the many political factors influencing the development of the socio-ecosystem. Unfortunately, the development of civilization is proceeding in a fashion whereby state frontiers and political and administrative borders do not coincide with the frontiers of ecosystems. Dozens of states close to the Black Sea region and the Danube have a bearing on the condition of the socio-ecosystem of the Black Sea. Each of them has its own particular economical and ideological interests that effectively determine the character of state policy. That is why each state approaches the ecological problems of the region on the basis of what it sees as being of relevance in its own part of the ecosystem. For most countries in the region protecting the environment has ceased to be a priority issue, in many cases merely serving as a smoke-screen behind which other problems are solved, an easy way of gaining a few privileges or a little help from the world community. Nowadays such a policy is actively condoned by public opinion. In Russia, for example, the population's interest in the environment has fallen over the past decade from 2nd or 3rd in its list of priorities to somewhere in the twenties.

Modifications in the geopolitical situation on the territory of the former USSR have led to changes in the dynamics of anthropogenic influence on the environment of the Black Sea basin both within Russia and in other CIS states and have created the preconditions for even greater change in the future. In combination with natural fluctuations and possible mutations of climate in the years ahead, this can become an obstacle to the development of the region's economy and greatly heightens the possibility of ecological crisis.

The fraught state of political relations in the Black Sea region is hindering the elaboration of a united programme of action. The many interstate accords and international programmes, including those conducted under the aegis of the European Council, have not been able to radically change the situation. The main reason for this is the lack of a complex, systematic approach to the region's problems. Without such an approach it will not be possible to stem the crisis and restore the balance of the ecosystem for another 15 years at least.

International cooperation can play an important role in providing ecological security for the Black Sea region. Although there are only six countries situated directly against the Black Sea (Russia, Georgia, Turkey, Bulgaria, Romania and the Ukraine), polluted substances from a total of 17 countries are swept into it if water flows from further afield are taken account of. In these circumstances, there is an urgent necessity for joint endeavours and the implementation of a common strategy on the environment as a means of coordinating internal legislation and state policy on nature conservation.

In 1992 the countries around the Black Sea signed the "Convention on the Protection of the Black Sea against Pollution" and additional protocols. A declaration on Black Sea economic cooperation and a Ministerial declaration on Black Sea conservation were also signed by the heads of these countries. In 1993, a three-year ecological programme on the Black Sea (BSEP) was elaborated at the World Ecological Fund with financial support from the World Ecological Fund itself, the then EC, Austria, Canada, Japan, the Netherlands and Norway. The coordinating centre for the Programme was in Istanbul, while centres for its practical effectuation were set up in each of the Black Sea countries. BSEP study groups prepared reports as a basis for the drawing up of a strategic plan of action geared towards restoring economic prosperity and ecological balance to the Black Sea.

The ecological prospects of the Black Sea depend in the first instance on the ability of the authorities in the Black Sea countries to realise that, given the nature of the Black Sea's ecosystem, ecological problems can only be solved by uniting efforts to improve the balance of the ecosystem as a whole, not just parts of it. It is precisely the ecosystem dimension that provides the objective reasons for cooperation, for using the ecological problems involved as a basis for uniting interests and actions. A classical example of this is the way all states bordering the Rhine cooperated in cleaning up the river's drainage basin. At the opposite end of the scale, because of a lack of coordinated efforts by the states around Aral Sea, degradation of the Sea's ecosystem has accelerated.

To prevent the Black Sea acquiring the status of a "dead sea", it is necessary to engineer a change in executive paradigms. The technocratic, antropocentric vision of the world must be replaced by a prioritisation of the ecological dimension, by a co-evolutionary approach to the interaction between humanity and Nature.

Practically, this may mean the creation of a supranational administrative body devoted to safeguarding the ecological balance in the Black Sea and with a remit to implement measures to improve it. The Black Sea states would need to surrender a number of functions to this body. A will to unite is called for that is able to harness the entire range of creative forces available. Having examined the perspectives of this problem, one can only hope that a "single Europe" type scenario will be forthcoming in the Black Sea region.

The Black Sea ecosystem is unique. This is expressed first and foremost in the following way. The ratio of surrounding drainage land to surface of water - 4,0 - is one order greater than for the world's oceans (0.4). This level of dependence on the situation on land serves as a specific index of the intensive influence of the anthropogenic factor on the ecosystem. In other words the Black Sea is excessively sensitive to human activity in its basin.

Today there are a wide range of factors impacting on the state of the Black Sea's ecosystem and affecting its balance. One could actually argue that the regulatory machinery has broken down and that the ecosystem is incapable of being returned to its normal state. The process of degradation, the transformation of the ecosystem into that of a "dead sea", is so far advanced that, unless urgent measures are taken, the inhabitants of the Black Sea states face ecological catastrophe within 30–40 years.

The major conduit for pollution is the atmosphere. The main causes of atmospheric pollution are coal combustion, oil products, and emissions from motor vehicles. Can this problem be solved? Already today, scientific-technical deliberations are answering this question in the affirmative. First of all it is necessary to create modes of power generation. One of its elements could be hydrogen power generation and the use of energy accumulator substances. Hydrogen power generation could become a panacea for most of the ecological problems in the Black Sea region. Use of hydrogen as fuel would solve the problem of pollution of the air, land and water in the region. The source hydrogen could be saturated with hydrogen sulphide water from the Black Sea. Extraction of hydrogen sulphide could in some measure stop or reduce the movement of the frontiers of the hydrogen sulphide zone. It is penetration of hydrogen sulphide into the surface water stratum that can lead to superior forms of marine life being destroyed.

Science cannot provide an unambiguous answer to the problem, cannot explain the mechanism of this phenomenon. Two possible sets of scenarios are frequently juxtaposed:

1. regulation and reduction of river flow, increasing salt content in surface of water, reduction of the flow of surface water through the Bosporus, and increasing inflow from the Sea of Marmara;
2. eutrophication of the sea, increase of organic mass reaching its depths and more intensive production of hydrogen sulphide. In anaerobic conditions, however, organic residue ferments instead of oxidising and the resultant hydrogen sulphide contaminates all around.

Nevertheless, measures to reduce levels of organic matter and fertilizers entering the sea would undoubtedly contribute to slowing the process down, as can increasing the flow of fresh water into the Sea of Azov/Black Sea basin to 10 % .

Is the threat of hydrogen sulphide real? Nobody knows the answer to this question. To the mind of some researchers the process is irreversible. Others regard it as an element of natural periodic fluctuations. A third school of thought argues that it is a purely anthropogenic phenomenon and everything depends on what is done by humanity.

Owing to the difficult economic situation of most of the Black Sea countries, pollution with industrial and agricultural effluents has decreased over recent years. For the past two years, waters of off-shore regions in the Russian section have been rated "moderately polluted", with the situation tending to improve. Over the next few years, however, purification plant will begin to break down, having had nothing invested in it for almost 10 years. Even now, mines in the Donbas region are throwing practically untreated mine effluent into the local water. Unfettered private construction in the health resorts has led to an excess of communal purification works. As a result, there has been a sharp increase in epidemics along the whole sea coast. If we do not take effective action soon, pollution will cause the chemical structure of the water to mutate and the consequences of human activity will make themselves felt throughout the biological and ecological chain.

As a result of "red inflow" and the mass destruction of organisms on the shelf, cherished species of fish have ceased to be traded. By the year 2000, bioproductivity is expected to fall by 25 % due to pollution by oil and oil-based products, by 30 % due to pollution by heavy metals, and by 25 % due to pollution by chemicals.

The integrity of Russia's food supply chain could be seriously threatened by a loss of marine bio-resources. Pollution of the Sea of Azov has depleted nutritional sources for the fish population. Unable to pass areas where treated effluent from Istanbul is ejected into the Black Sea, some types of marmara fish have disappeared.

The most acute problem is marine pollution by heavy metals and by their constituent parts. In nature there is no mechanism for self-purification in the face of such an onslaught. Passing through the food chain, they eventually have a disastrous effect on humans. Unfortunately, the problem of the extent to which heavy metals harm molluscs and fish in the Black Sea is not considered an acute one. However experience gained with other bodies of water indicates there is indeed a danger. At the end of 80s, outbreaks of sturgeon disease were identified in the Volga and Kaspi. It appears that this was a result of the immunity of fish having been reduced through pollution by heavy metals.

At the present time a process of establishing a new economic model for the region is in train. New factors bearing on the ecological stability of the basin are materialising as a result. Among these are the widening of existing seaports and construction of new ones, increases in import volumes and oil exports by sea, changes in the structure of energy consumption in the countries of the region, plans by Ukraine, Bulgaria, and Romania to tap oil and gas from the sea shelf, and the destruction of naval installations and ageing of the navy's ships.

A considerable influence on the ecological condition of the water in the shelf zone is being exercised by soil erosion, and also by the harvesting of Don molluscs using dragnets in the Don trawling zone. Over the past 30 years, 21.6 million cubic metres of soil have been deposited onto the dust heaps of the Kerch strait alone. With most cleaning vessels tied up, there remains the problem of clearing up the channels of ports. Observations show that over recent years dangerous substances have increasingly been ending up in the water during unloading at sea and river ports. Up until recently, the volume and type of dangerous substances conveyed was relatively low. Now the size and number of such cargoes being transferred within or through the Black Sea basin have significantly risen.

The most acute type of pollution is that involving oil and oil products. It is known, that for one hectare of sea surface covered by a film of oil, more than 100 million fish grubs and a great deal of other sea organisms perish, sharply lowering the exchange of oxygen. Today the total area of oil slicks is 0.9 % of the total area of the Black Sea. In 1993, 8.5 thousand tonnes of oil products were jettisoned in the Russian section of the Black Sea alone.

What really sets the alarm bells ringing are plans to turn the Black Sea into one of the main transit routes for Caspian and Middle East oil bound for Europe, not to mention the prospects of oil rigs on the Black Sea shelf. It is immaterial to Nature whether transportation be by tanker or by some form of pipeline. What does matter is that the danger of ecological risk is high in both cases. Turkey has already taken all due measures, having prohibited tankers from passing through its straits.

In these projects the opinion of ecologists must be taken into consideration from the outset. But while we see the economic, political and even ideological aspects of projects discussed, one cannot sense any concern on the part of local authorities or of populations in the regions concerning the environmental impact. There are no certainly no expert assessments being carried out on of the ecological safety of projects. World practice shows that the appalling economic damage incurred when a supertanker founders or a pipeline is breached will bring to nought the possible economic effect to be had from oil transportation. It is worth bearing in mind that any benefit to the oil producers and transport workers will be more than offset by what shore-line regions will have to pay for in terms of pollution. The cost of damage to the ecosystem in the wake of an oil spill is put at US $ 250–2,500 per tonne of spillage depending on the type of ecosystem.

Of course, not all ecological phenomena can be quantified. A wide variety of interactions in marine ecosystems are not studied in full, while the complex nature of the ecological consequences of human actions upon the environment limits the extent to which public and social losses resulting from degradation of the marine environment can be accurately assessed.

There is no conclusive proof of radioactive contamination of the Black Sea, for instance in the aftermath of the disaster at the atomic station in Chernobyl. The construction of an intricate system for monitoring the state of the environment is envisaged as a combined effort by all the Black Sea countries. For the present we are producing fragmentary and often unreliable ecological information. The absence of exhaustive unambiguous characteristics in the current ecological condition of the region means that decisions are having to be made within a framework of uncertainty. It is for this reason that evolved nature conservation measures are not always applicable in the present situation.

It is especially difficult to estimate the indirect, long-term consequences of pollution, those that will not make themselves felt for perhaps several generations. The problems of the Black Sea region require coordinated efforts not only on the part of scientists from the countries of the region but also of the international scientific community. An ecosystem approach to the problems of the Black Sea could form the basis of an ecological culture and ecological world outlook for more than the 160 million or so people living in the Black Sea basin.

Let us pause to consider another problem. Mass fishing with fine-mesh cellular nets has led to the genetic potential of some kinds of fish being seriously impaired. Species have begun reproducing smaller offspring, virtually new subspecies. A similar problem has emerged in fish farming. Whereas, in nature, preeminence in reproduction is accorded the more viable genus, in the course of artificial reproduction this is impossible to achieve. At the genetic level the mechanisms of adaptability and viability mutate.

The Black Sea comprises semirelict reservoirs isolated even from Mediterranean waters. That is why new arrivals, meeting no serious competition on the part of local fauna, are able to flourish. At the beginning of the 80s, mnemiopsis leidyi was introduced via the bilge water of wheat-cargo ships from the USA. Within 10 years bays and shallows near the shore had been colonised by these animals. Mnemiopsis leidyi is a predator. Appearing in huge concentrations, mnemiopsis leidyi, which devours organisms or zooplankton ranging from infusoria to fish grubs, could not rely for long on the structure of plankton associations and on its fish stock. Of course, having exhausted its supply of nutrition, the mneopsis leidyi will very rapidly dwindle in number. But self-restoration of biodiversity even to recent levels will not take place. The ecosystem needs human help.

The methodological basis for best solving the Black Sea's ecological dilemma can usefully be the concept of sustainable development adopted by the international community at the UN Conference in Rio de Janeiro. In Russia this model of development is legislatively anchored as "Conception of Transition of the Russian Federation to Sustainable Development", adopted by Decree of the President of the RF in April 1996. The elaboration at state level of a concept of sustainable development represents a transition towards implementing ideas that are engaging the minds of civilized humanity.

At the present time, the Government of Russia is working on a strategy of sustainable development the foundations of which comprise the principles of 'green' economics and a systematically ecological approach to development in the regions. The ecosystems of the Caspian Sea and Sea of Azov-Black Sea basin have the potential to become the testing ground on which to apply the principles and main ideas of sustainable development. After all, the ecological interests of all states in the region are virtually identical.

First of all it is necessary to halt degradation of the ecosystems forming the marine and general environment along their coasts. To this end, it is necessary to pursue the following goals: re-instating environmentally fertile landscape functions; creating the preconditions for organic renewal of natural resources and, most important of all, of healthy water in small rivers; restoring biological productivity and the specific variety of marine fauna and flora; considerably reducing levels of biogenic combinations, chemical pollution, and weighted substances in zones of water collection and in the main channels; recultivating the coast zone and, most crucially, seashores; making optimum use of the balneological resources of the sea. No less than 10 years will be required to achieve these objectives, even given a favourable constellation of circumstances. Taking into consideration the peculiarities of spatial organisation in nature as well as the socio-economic and political subsystems of the Black Sea region, ecological problems are likely to be solved here at differing speeds, which is why they will appear to be solved as a unit less quickly.

Improving the ecological situation in river basins requires substantial financial and material expenditure over a vast area. Such processes are characterised by considerable inertia, which is why ecological adaptation of nature use in river basins will not induce positive changes in the sea for a very long time to come. On the seashore, negative anthropogenic influences on the environment impact more rapidly on the condition of the marine ecosystem. Here, the concrete sources of pollution can be rapidly located, measures to eliminate them drawn up, and the desired effect rapidly attained. In the waters of the Black Sea, by contrast, any enhancement of the ecological situation involves observance of nature conservation norms and rules by shipping as well as in trade with sea products and other human economic activity. Monitoring observance could become the most effective tool with which to improve the natural marine environment.

Should the problem of improving the ecosystem of the Black Sea not be solved within the next ten years, we will be confronted with a deteriorated human environment, depleted seafood resources, losses through reductions in the number of tourists and guests, and losses suffered by hunters and sportspeople. And who can estimate how the aesthetic qualities of the marine environment will be affected? Civilization will suffer not only material, but moral damage, which in most cases cannot be quantified. It is necessary to instil into the consciousness of everyone the bounds of water's propensity for self-purification in the off-shore zone before addressing the necessity of urgently eliminating the damage, restoring the lost balance, and conserving this unique ecosystem for the future. It is this factor towards which the planning of sustainable regional development must be orientated.

After the disintegration of the USSR, the Black Sea region became for Russia the focus of many of its interests - economical, communicational, recreational, ecological. A new policy of relationships between the region's states needs to be elaborated. Russia has ratified the "Convention on the Protection of the Black Sea from Pollution" together with further international policies designed to regulate cooperative action by states within the Black Sea's ecosystem. To coordinate the work of ministers and departments in the Russian Federation, in 1993 the Government of Russia set up an interdepartmental Commission for the Black Sea. Decisions of the Commission are binding on all bodies represented in it, as they are for enterprises and organisations operating in the Black Sea region. The Commission is paying great attention to issues such as assessing the condition of the marine ecosystem and marine biological resources, the main watercourses and reservoirs, and protecting the Black Sea and the Sea of Azov from pollution originating on land and from activities on the sea shelf. The issues of waste management, improving navigation, developing the system of monitoring the environment, and conducting coast-strengthening works have also been considered.

Geographical Information Systems in Regional Planning and Management for the Ukrainian Black Sea Coast[1]

Vladimir N. Anderson
GIS Laboratory
Department of Economic and Social Geography, Odessa State University
Odessa, Ukraine

Introduction

Ukraine as a new independent state needs a new methodology and new technologies in the sphere of regional planning and management if it is to plug into the western information-based planning culture in the drive to solve its complex, deeply-rooted economic, social and environmental problems.

Regional planning and management, by its very nature as an information-rich discipline, has much to gain from the facilities which the new information technology provides (Stillwell and Scholten, 1990). As the authors have pointed out, "In different countries, the activities which comprise procedural and substantive planning have evolved over time, yet the basic ingredients of formulating hypotheses based on problem analysis, of identifying goals and objectives, of designing alternative strategies and formulating new policies, of implementing plans and programmes, and of monitoring and updating, remain intact. The effectiveness with which any one of these process activities is undertaken depends upon the availability and appropriate utilisation of information of different types which is obtained from various sources. Information organised within Geographical Information Systems (GIS) therefore provides a framework to support the processes of decision making which together constitute planning practice" (p XXIII).

At the Department of Economic and Social Geography at Odessa State University we are involved in designing a multipurpose GIS of the Ukrainian Black Sea Coast for regional planning and integrated management. On the basis of an integral expert assessment of the automated information environment, GIS has the

[1] This work has been completed with generous support from the Faculty of Geographical Sciences at Utrecht University, Netherlands. The author would like to personally thank Prof. Henk F. L. Ottens and Dr. Jan Jaap Harts for their invaluable contribution towards the setting-up of GIS teaching and research facilities at Odessa State University.

task of providing in-depth planning scrutiny and monitoring of economic and social entities and supporting in this way the planner's decision making process.

This paper sets out the key methodological, methodical, legislative and practical issues involved in GIS implementation in regional planning and management, taking as its example the new Ukrainian planning macroregion.

1 Regional/Spatial Planning in Ukraine: From Soviet Tradition to a Europe-wide Perspective

1.1 Regions and Regionalism: the Case of Ukraine

Nowadays differing political, scientific and legal orientations are to be discerned in the Ukrainian State's regional policy concepts. On the one hand, the Ukrainian system of regions is a result of the great diversity of natural conditions within the country and of historical differences in the economic development of the various parts of Ukraine. There are regions that have changed their statehood several times and in the most dramatic fashion. Among these are left-bank and right-bank Ukraine, Slobozhanshchina, Getmanshchina, Donbass, the TransDnieper region, Polesye, Volyn, Podolye, the Carpathian region (Galichina), Bukovina, the Trans-Carpathian region, the Ukrainian section of the Black Sea coast, and the Crimea. Considerable differences in the natural conditions and resources of these regions, different histories of settlement and economic utilisation, various sets of development problems – economic, socio-demographic, ethnocultural, ecological – emphasize the need for regional approaches and policy and are used as arguments in favour of a federal state system for Ukraine.

On the other hand, such diversity of political history, natural economy, and ethnic nationality provides pressing reasons for preservation of the current unitary Ukrainian state. The orientation towards a federal system with its inevitable centrifugal tendencies seems dangerous now given that the country has just six years experience of statehood and given the fledgeling nature of its formal machinery.

The principles and mechanics of state regional policy are being actively developed in Ukraine at the present time. Forms of decentralised management and interregional and, indeed, international collaboration are being piloted in the regions. Some of the Ukrainian regions have already mapped out strategies for their socio-economic development. Such conceptions for two regions – the Carpathian and Ukrainian Black Sea coast – are being examined at government level, and elaboration of state programmes on socio-economic development for these regions has begun in accordance with decrees by the Ukrainian government (1995).

The strength of 'regionalism' as a movement is strikingly exemplified by the international cooperation of Ukrainian regions. The Carpathian region, for example, is involved in the international 'Eurocarpathian region' project, the Volynskaja and Rivnenskaya regions in the 'Europe-Bug' project. The Ukrainian Black Sea coast region is representing Ukraine in international programmes and projects relating to integrated coastal management in the Sea of Azov-Black Sea basin, the economic development of littoral zones (in particular, use of their oil and gas resources), and economic integration of the states around the Black Sea. Moreover, the Odessa region is the Ukrainian representative in the Association of European Regions.

The blueprint of Ukrainian regionalism has to be elaborated on the basis of a problem-oriented approach within the framework of a sustainable development paradigm: for each region the main task of socio-economic development and environment preservation has to be defined in a way that limits and regulates the development of many economic branches and activities as well as giving a common vector for economic/environmental policy making. Let us briefly shed light on some of the main problems facing the Ukrainian regions.

For the *Carpathian region* the major task involves preservation and rational use of the Carpathians with their unique biosphere potential, the latter playing a major role with respect to environmental protection not only in the regional but also in the European and even global context. The issues involved are not only the recreational and tourist amenities of the Ukrainian Carpathians but also their global-biosphere functions – ensuring water resources, influencing climate, sustaining the environment.

In *Donbass* there are major interconnected problems relating to the water supply and environmental degradation that have yet to be tackled. Mass devastation of natural hydro-geological regimes by mining has destroyed the natural water regeneration cycle in the region and caused problems with the water supply and in particular the quality of drinking water, which is now provided from other regions. Chaotic urbanisation associated with the mining industry, a deteriorating environmental situation across the board, intensive pollution of underground and surface water, accumulation of hazardous waste – these all demand urgent action, the erection of effective protected-hygiene zones, industrial relocation, and a more rational spatial organisation of the economic complex as a whole.

For the *Polesye region* the dominant problem is the conservation of marshy lands which serve as a powerful natural regulator, surface filter, and underground drain for a great part of the Ukrainian land mass and which are being subjected to intense strain from a massive drainage melioration programme.

The Ukrainian *Black Sea coast region* (Odessa, Nikolayev, Kherson oblasts and the Crimea) is primarily concerned with the problem of rational use of seaside areas. World experience testifies convincingly to the exceptional role of coastal zones, which have to be considered as a unique, polyfunctional, highly valuable national resource. In Ukraine, coastal zones and areas enjoy high ratings both amongst the population and in the economic domain. They feature very high concentrations of habitation, industry, and infrastructure as well as production assets and capacities, all far in excess of comparable figures for more inland regions. At the same time the coastal zones with their natural estuary complexes, sea coast and shelf zones fulfil important biosphere functions (regional and global). This is where the unique natural complexes and biocoenoses are concentrated, seasonal birds migrate and sea fauna routes cross, many of which have the status of most protected objects and territories. It is precisely the sea coast, estuary areas of rivers, and the adjoining shelf habitats that are the most vulnerable ecologically.

As we can see, strong ontological arguments exist in favour of regionalism in Ukraine. But how can these be translated into concepts of regional/spatial planning and state legislature? First we shall explore the latter issue.

1.2 Searching for a Ukrainian Model of State Regional Policy

Regionalism as a state policy has to be a system-forming determinant which not only influences the direction Ukrainian statehood takes but is also an effective aid to the country's move towards democracy, stability and sustainable development. It is now at the stage of being an issue of practical research: the tasks have been identified but have yet to be lent concrete shape and there is as yet no means of carrying them out.

There are currently processes of privatisation of property, decentralisation of economic management, and the upgrading of local self-government going on in Ukraine. They bring with them the need to develop and implement the unified principles and methods of economic, social, demographic, ethnic and environmental policies in regional planning. Though regional policy needs to be coordinated at interregional and national levels, real and effective planning/management of socio-economic processes is possible only at the regional level. Ignoring the economic, social-demographic, ethnic and other regional peculiarities of the 'territorial collectives' leads to the growth of social pressures and tension, indeed the absence of state regional policy actually leads to the regions demanding more of the centre.

A constructive approach to the problem presupposes analysis of the concrete historical conditions of state formation in Ukraine and in its separate regions. The main thing here is to provide a balanced development of the regions, taking into consideration their specific features and orientations whilst preventing a potential 'balkanisation' or 'lebanonisation'. Therefore it is necessary to consider not only the concrete socio-economic realities of post-Soviet Ukraine, its history in retrospect, the mentality of the population and their willingness and ability to participate in the process of reforming state system, but also the rather contradictory tendencies in theory and practice of young countries in general. On the one hand, it was a federalist 'revolution' that gripped the whole world and has been accompanied by a reassessment of federalist ideas concerning the real subject contained in unitarist/federalist/confederalist/constitutional decentralisation paradigms. On the other hand, there is a sufficiently widespread prognostic notion about the coming of an era when public power will be divided between communes, regions, national and European states.

It is obvious that surmounting the social pressures of the 'external' influences being exerted upon the process of Ukrainian statehood make it impossible to meet demands for greater independence in diverse spheres of regional life, or agree to autonomy for the various territories or to federalisation of the country. The solution to the problem lies in postulating and implementing regional policy legislation that is, however, largely alien to the historical experience of Ukraine and cannot necessarily be adopted on the strength of the models which foreign experts, politicians and scientists insist on recommending. It is necessary for Ukraine to find its own model of state, conceptually geared to genuine regionalisation of the country, and hence supporting the drive towards a decentralised constitutional state.

1.3 Post-Soviet Regional/Spatial Planning Paradigm: the Ukrainian Point of View

To discuss various regional/spatial planning paradigms relevant to the post-Soviet 'space-time' we have first to identify the core concepts of regional planning *per se*.

As it was proclaimed in the European Regional/Spatial Planning Charter (Torremolinos, 20 May 1983), 'regional/spatial planning gives geographical expression to the economic, social, cultural and ecological policies of society' and 'contributes to a better spatial organisation in Europe and to the finding of solutions for problems which go beyond the national framework and thus aims to create a feeling of common identity...' Regional/spatial planning should be democratic (it should be conducted in such a way as to ensure the participation of the people concerned and their political representatives), comprehensive (it should ensure the coordination of the various sectoral policies and integrate them in an overall approach), functional (it needs to take account of the existence of regional consciousness based on common values, culture and interests sometimes crossing administrative and territorial boundaries), long-term oriented (it should analyse and take into consideration the long-term trends and developments of economic, social, cultural, ecological and environmental phenomena).

A felicitous formulation relating to the application of new information technologies in regional/spatial planning was provided by Ottens (1990):

> The term 'planning' covers a wide range of activities. In the planning literature, strategic planning is often distinguished from operational planning (...) Strategic planning has to do with long and medium term decision making. It often involves a lot of research, discussions, consultations and negotiations. The activities that support decision making can be split into two groups: the organisation of the decision making process itself and the production of tangible results, in the form of plans, programmes and project initiatives. These two aspects of strategic planning are often referred to as procedural and substantive planning (...) Procedural planning produces the organisational and decision making infrastructure in which substantive planners can produce research reports, policy reports, information material and, eventually, official plans, plan revisions, implementation programmes and project outlines. Initia-

> tives and control activities connected with the implementation of plans that are in force are called operational planning or action-oriented planning (...) This planning aspect involves the judgement of applications and the issuing of permits with respect to development, building and installation. But it can also include the monitoring and control of projects in progress. At higher tiers of government, the evaluation and approval of planning activities performed by lower governmental bodies is part of operational planning. (p 15)

As a matter of fact, the Soviet state regional planning paradigm was designed to serve a highly monopolised empire and had strong sectoral and exterritorial aspects. Despite the officially proclaimed policy of the 'balanced socio-economic development of the regions', the interests of the 'unitary Soviet state' were paramount. Owing to the all-pervasive 'economy of shortage', it was not possible to improve the quality of life in respect of goods and food consumption at regional level. The sectoral approach in economic development led to an aggravation of the environment and exhaustion of natural resources. The hierarchy of regional socio-economic management (national, regional, local) was not well defined and it prevented coordination and cooperation between the various levels of decision-making and the balancing of financial resources. The local and regional authorities involved in regional/spatial planning policy did not have the power to take and carry out decisions, nor adequate financial means. And it was not possible for the citizen to participate in policy making for regional/spatial planning.

Unfortunately, the bottle-necks of regional planning noted above are still palpable in post-Soviet Ukraine. An additional factor is having a strong and progressively negative influence on the national economy and regional policy in the new independent state – a financial crisis in which the state cannot continue funding its sectoral economies and activities due to a serious budget deficit. This has already caused the degradation of some Ukrainian regions and resulted in dangerous demographic, migratory, and socio-political consequences and trends. In such conditions, a well-founded and balanced regional policy may be viewed as an important prerequisite for improvement of the socio-economic and ecological situation in Ukraine. Ukrainian experts depart from the assumption that approximately 60 % of state investments have to be allotted to regional socio-economic programmes aiming to achieve 'territorial justice' in the state (Pistun, 1996).

The main and most effective way of implementing long-term regional policy in Ukraine is argued to be to elaborate regional programmes of socio-economic development for individual oblasts and rayons. First of all such programmes have to locate productive forces in economically depressed rayons as well as in highly urbanised industrial regions and large cities (so as to make more complete use of the existing socio-economic potential). And regional programmes need to be used not only to solve regional problems and rationally organise the economy by area, but also as a tool for fostering socio-economic development as a whole.

The main practical issues facing regional planning in the emerging post-Soviet Ukrainian can be summarised as follows (Pistun, 1996):

I. quantitative and qualitative assessment of integral resource potential – natural, labour-related, financial, recreational, infrastructural, or relating to innovative activity, investment capacity etc. – by all territorial units defined in the context of regional planning: It foresees elaboration of a set of norms on resource usage which have to be obligatory at all levels of regional planning and management.
II. analysis of the geodemographic situation and trends is being conducted in the regions, i.e. assessment of existing settlement systems and their social infrastructure. The demographic situation in Ukraine now is rather complicated and unstable because there are many economic, political and geographical factors acting on the population. Therefore it is necessary to design the system of geodemographic monitoring using the concept of 'population quality' as a base.
III. improvement of the complex proportionality of socio-economic development in the individual economic regions, oblasts (regions), large cities and urban agglomerations as well as reconstruction of the old industrial rayons and territories. Complex regional planning has to deliver development and territorial organisation for all macroregions, economic regions, free economic zones, and urban centres of the future 'lands' (mooted administrative territorial units in Ukraine).
IV. underpinning of the territorial proportionality of state development by means of socio-economic regionalisation on the basis of empirical geographical data and spatial analysis. It is necessary to use two basic schemes of socio-economic regionalisation in regional planning: first one has to register the current spatial structure of society, then one has to determine its prospective patterns. Prospective regions also have to be viewed as the objects of state management.
V. updating of administrative territorial divisions of Ukraine to improve the state regional management system by means of optimal combination of national and local interests.
VI. analysis and effective usage of the socio-economic and cultural potential of the new geopolitical position of Ukraine. This provides assessment of regions and territories with the purpose of creating free economic zones and joint ventures in frontier areas.
VII. environmental assessment of the territory and ecological scrutiny of all regional projects and programmes underway in the state, ecological regionalisation of Ukraine, elaboration of national ecological insurance system to be implemented in regional planning and management.

2 Geographical Information Systems: New Technology of Regional/Spatial Planning and Management

2.1 Geographical Information in Regional Planning and Policy Making

Regional planning and regional policy-making involve a variety of geographical information at different spatial (national, regional, local) and time (long, medium, short term decision making) levels. In the different phases and for different activities of the planning process different geographical data are required and used. To follow Ottens's description of the nature of regional planning mentioned above, it is useful to classify the geographical information relevant to planning by making use of three independent (orthogonal) 'axes':

X - 'Term of decision-making' (e.g. strategic or operational planning);
Y - 'Spatial level' (e.g. national, macroregional, regional or local);
Z - 'Planning objective' (e.g. physical or socio-economic planning).

Of course, in every 'cell' of this classification we will deal with specific needs and requirements. For example, the elaboration of aggregate national physical plans or socio-economic programmes contrasts with the processes of development monitoring and land use allocation in small areas. As physical planning focuses on conflicts which arise because of the increasing demand for space from competing economic and social interests, it needs geographical information on agriculture, housing, manufacturing, services, public infrastructure, and recreation presented on various large-scale thematic maps and charts. Strategic macroregional planning, by contrast, calls for rather generalised socio-economic geographical data in the form of integral small-scale maps. The geographical information used in decision-making in the course of regional policy making may be of various types, extensive in quantity, variable in quality and applying to areal units of different size (e.g. in Ukraine the following administrative territorial units are often used: local Soviets, towns, rayons, cities, oblasts, macroregions). In the course of operational planning planners need current geographical information for monitoring and control of projects in development. Such information is sometimes only available by means of remote sensing and necessitates the deployment of modern telecommunications.

Paradoxically, it is a fact that despite large amounts of so called 'official' or 'conventional' cartographic (and non-cartographic) presentations of geographical planning data, decisions come to planners via their 'mental' maps. So, the task consists in creating an information environment so favourable that it gives planners the most robust and complete mental geographical image of the planning problem conceivable. As is argued in the next paragraph, GIS technology provides decision makers with an automated information environment in which geographical data are presented in an effective and attractive fashion.

2.2 GIS Dimensions of Regional/Spatial Planning and Management

What is a GIS? The most widely quoted definition describes GIS as 'a powerful set of tools for collecting, storing, retrieving at will, transforming and displaying spatial data from the real world' (Burrough, 1983). A GIS has to solve three main tasks:

1. storage, management and integration of large amounts of spatially referenced data;
2. providing the means to carry out analyses which relate specifically to the geographic component of the data;
3. organisation and management of large quantities of data in such a way that the information is easily accessible to all users (Scholten and Stillwell, 1990).

Regional planning and management is a promising arena for the practical application of GIS technology. This is precisely where the main applied function of GIS – to transform spatially distributed data into planning and management decisions – is graphically demonstrated. As is known, the final aim of the use of any information system is acceptance of a decision. In that sense geographical information systems are an extraordinarily useful means of supporting spatial decisions. The reality of the situation is, as world experience testifies, that GISs are used more frequently as a spatial decision aid in such important applied fields as land use planning and management, infrastructure development, nature resource use, geographical marketing and geodemography. GIS accumulates, analyses, and visualises the necessary geographical information. With regard to the latter functions (analysis and visualisation), GIS technology has similarities with the technology of Decision Support Systems (DSS).

DSS comprises a wide range of information technologies capable of providing factual, logical, methodical and heuristic assistance and of solving awkward, badly defined problems. Such problems are not so unique that special methodical apparatus must be elaborated for them, and at the same time they are not so typical that they can be adapted to already existing strict schemes of decision-making. It should be noted that any computer system, beginning with the simplest Database Management Systems (DBMS) and ending with the program packages of mathematical modelling and optimisation, can 'render support' in the process of decision making. However, in the case of spatial decisions it is necessary to be able to integrate the various spatially distributed data over the geographical environment and present them to users in the form of computer maps and atlases. Although the analytical abilities of the modern GIS are limited, their combination with mathematical modelling and technologies of artificial intelligence (expert and consulting systems) generates an absolutely new kind of information technology – the Spatial Decision Support Systems (SDSS). Because of the presence of a functional connection between spatial data and spatial models in these systems it is possible to create an integrated informal analytic environment that will facilitate acceptance of difficult spatial decisions.

GIS technology has caused a revolution in thematic mapping for spatial planning because it makes use of and delivers computer graphics (3D-mapping and animation), interactivity (obtaining data directly from the map, real time calculations, rich cartographic design), multi- and hypermedia (integration of maps, text, videos, sound and animation pictures). The media-integrated GIS enables the planner not only to collect, analyse and present attributive data as overlays in the thematic digital map but to act as an author of planning hyper documents (Kiib and Veirum, 1993). Moreover, such technology is eminently conducive to a comprehensive mental vision of the planning problem by decision makers.

2.3 GIS-based Regional Planning in Ukraine

In Ukraine as in the other new independent states there is now a process underway of learning from western GIS experience (mostly in academic and commercial spheres). Post-Soviet regional planning culture is still in the phase of 'promising emergence' and the national experience of GIS-based planning and management has a rather pioneering and adaptational character.

The first planning-oriented GISs were applied to urban planning and management for a number of large Ukrainian cities at the beginning of the 1990s. Despite the fact that they have had mainly a specialised cadastral function, their appearance has given a boost to the application of new information technology in state urban planning institutions and agencies. In Kiev, for example, there already exist several specialist organisations devoted to applying the GIS-based technology of urban spatial planning and management in four basic directions:

1. design and implementation of urban cadastral systems;
2. topographic and land-surveying service on the basis of digital terrain modelling;
3. elaboration and construction of automated databases for the urban economy;
4. urban physical planning and design.

At the regional level some interesting GIS-projects have been initiated over the past few years by regional state administrations (regional representatives of the Presidency in Ukraine) and regional governments. In 1993 a regional information system was designed in Chernigov oblast to support decision-making in the field of socio-economic planning and management. It includes 28 ‘subsystems’ dealing with various sectors of regional economy and the social sphere. The most important of these are territorial organisation of the productive forces, the quality of life of the population, budget and finances, resources, ecology and nature preservation, industrial infrastructure, services and land use. The system was designed on an IBM PC platform using commercial GIS software (ARC/INFO). Digital basemaps of the regional GIS were produced from original paper maps of 1 : 200 000 and 1 : 400 000 scale.

A large and important GIS-project is now being set up in the Autonomous Republic of Crimea. In 1995 regional government adopted a special programme on elaboration and implementation of the United Republican Digital Territorial Cadastre. It is envisaged as a kind of regional GIS to provide data back-up for management decisions arrived at by the executive and relating to the utilisation of the Crimea's natural resource potential.

At geographical departments in some Ukrainian universities, special courses on GIS and new planning-oriented specialist disciplines have been introduced (e.g. 'Management of Regional Development' at Kiev university, 'Urban and Regional Economics' at the Kharkov urban economy academy, or 'Geo-ecology' at Odessa university). This shows that in the Ukrainian academic sphere there appears to be a strong drive towards sharing western experience in the field of regional/spatial planning and management with the emphasis on modern information technologies, and the familiar epithet 'state of the art' acquires here a rather extended meaning.

3 Developing Multipurpose GIS for the Ukrainian Black Sea Coast

3.1 The Ukrainian Black Sea Coast as a Planning Region

The Ukrainian Black Sea coast occupies an important geopolitical position and an especially advantageous position in transport geography terms, situated as it is at the confluence of waterways running from Central Europe to the Near East and South Eastern Asia and from the CIS states, notably Russia, to the countries of the Mediterranean basin and the great oceans. The region has a developed transport infrastructure, making for considerable distribution and transit potential at interregional and international levels. Placing the economic situation in the region on a sound footing has always required a universalist approach. The concept of regional development for the Ukrainian Black Sea Coast, uniting the efforts of the Odessa, Nikolayev, Kherson oblasts and of the Republic of Crimea in the fulfilment of common tasks, addresses itself to the key factors influencing regional economics. Strategically the concept has the task of coordinating the interests of the state and of the regions in creating conditions for a rational utilisation of local resources and the specific features of natural and productive capacities with the aim of achieving the most efficient form of socio-economic development. Consequently it is very important to create a mechanism of close interaction between the central and local bodies of the state executive and to transfer additional authority over implementation of economic reforms to regional level so as to further the cause of stable socio-economic structures in the regions. The regional state administrations in the oblasts, for instance, provide incentives for businesspeople to manufacture the products most needed in their particular region, offering a flexible fiscal regime for one thing. The Regional Programme for Fostering Business Activity has been approved and is being put to effect. With its help, the non-state share of total industrial production has grown considerably. At oblast level, a programme of demonopolisation has been elaborated, under the terms of which monopolist enterprises have been pinpointed. This is encouraging the development of various collective and individual types of property and stimulating local business activities where significant reserves are in place.

In the case of southern Ukraine we have a very jumbled and complex ethnic situation which has become more complicated still since 1991 when Ukraine obtained its state independence, the new geopolitical circumstances having left their mark on traditional ethnosocial processes in the region. As contemporary history shows, the ethnic factor has a strong and often unforeseen and unexpected influence on the economic and political development of the state. In many cases ethnic conflict and tension have been the main cause of political instability or even civil war, as witnessed by the examples of Yugoslavia, the Chechen Republic or Moldova.

Figure 1. The Ukrainian Black Sea Coast planning macroregion

Historically the southern Ukraine is a region where more than 40 ethnic groups and nations have resided. The largest groups among them are Russians, Bulgarians, Moldavians, Tartars, Jews, Gagausians, Belorussians, Armenians, Poles, Germans, Romanies, Greeks, Georgians, Czechs and Albanians. Most of them have been grouped in comparatively confined areas in the region as well as having been subjected to various forms of ethnopolitical discrimination (e.g. deportation of the Crimean Tartars, Germans and Bulgarians in the Soviet Union from 1937–

1944). From 1950–1980 the main influence on the ethnic structure of the population in the region took the form of migrations within the framework of the 'centrally planned' demographic policy that prevailed in the Soviet State during that period.

In May 1995 the Ukrainian Cabinet of Ministers adopted the State Programme of Socio-economic Development for the Ukrainian Black Sea Coast. In effect, a new planning region has been proclaimed in Ukraine, a region where ethnic diversity has to be taken into consideration as a dominant criterion in regional planning and policy making.

In practice state management of the region has to be effected by the regional state administrations of the oblasts and the Crimea. Elaboration of the State Programme was entrusted to the Odessa Institute for Market Problems and Economic/Ecological Studies, whilst its practical implementation is the responsibility of the Academy of Public Administration (Odessa Regional Branch). But owing to the fraught financial situation work has almost come to a standstill. Another problem concerns poor up-to-date information on ethnodemographic and ethnosocial processes in the region. Official census statistics are still unreliable as regards ethnosocial phenomena and do not take account of the major ethnosocial features at all. It is therefore necessary to draw up and dimension special composite indices in the form of ethnodemographic, ethno-economic, ethnocultural, ethnopolitical 'portraits' (or 'pictures') of all ethnic groups residing in the region. These will assist implementation of ethnosocial monitoring in a territorial and historical context by the regional state administrations.

Additional ethnogeographical research is needed in the region to analyse the processes of relocating displaced ethnic groups and nations (Crimean Tartars, Germans) as well as refugees from NIS countries. It is necessary to establish contacts between the local authorities and the heads of the national and cultural associations to involve them in research on the latest issues of regional development.

Ethnic socio-economic processes in the region remain an unexplored area. Profound changes in the economic structure of 'post-Soviet' society have caused extensive ethnic migration in the region that has been accompanied by ethnic conflict and tension. The ethnopolitical situation in the Crimea remains very unstable and unpredictable, largely owing to the freak ethnodemographic situation (share of population: Russian 57 %, Ukrainian 30 %, others 13 %). To enable this tension to be defused, a study of ethnopolitical processes and the mechanisms of social 'ethnopolitisation' is needed. For successful implementation of the State Programme on socio-economic development in the region it is necessary that each of its aspects be adapted to the specifics of all nationally autonomous groupings. A significant problem is posed by the extreme complexity and complementarity of this project, which includes research, managerial and legal dimensions and requires a single source of funding that is as yet unavailable in Ukraine.

3.2 ICM's View of the Ukrainian Black Sea Coast

Coastal areas throughout the world have historically been among the most heavily exploited areas because of their specific resources formed by the interactions and interrelationships between the land and the sea. In coastal countries today, an estimated half of the total population lives in coastal zones, and migration from inland areas to the coast is increasing. There is also sharp competition between the need for immediate consumption or use of coastal resources and the need to ensure the long-term availability of those resources. In many countries, this conflict has already reached a critical stage, with considerable parts of the coastal zone polluted from local or upland sources, the attraction of international and domestic tourism reduced by degraded environments and services, fisheries severely depleted or destroyed. If coastal resources are to be sustainably developed and coastal environments protected and restored, effective action is urgently needed. To answer this need, a customised long-term international program dubbed Integrated Coastal Management (ICM) has been devised.

ICM is a governmental process and comprises the legal and institutional framework necessary to ensure that development and management plans for coastal areas are integrated with environmental (including social) goals and are drawn up with the participation of those affected. The purpose of ICM is to maximise the benefits provided by the coastal zone and to minimise conflicts between individual activities and the harmful effects of one upon the other. ICM starts with an analytical process to set objectives for the development and management of the coastal areas. What ICM should ensure is that the process of setting objectives, planning and implementation involves as broad a spectrum of interest groups as possible, that the best possible compromise between the different interests is found, and that a balance is achieved in the overall use of the coastal area. The planning and management process must be comprehensive and take account of all the sectoral activities which affect the coastal area, its resources and environments. Therefore, the planning and management process must address economic and social issues as well as environmental and ecological concerns. The final goal of ICM is to harmonise these activities with other frequently non-sustainable and non-rational uses.

For the Ukrainian Black Sea Coast the main problem is the rational use of seaside zones and areas. World experience testifies compellingly to the exceptional role of seaside zones, which are considered as unique, polyfunctional, highly valuable national resources. The seaside zones and areas in Ukraine (with some exceptions) are attractive amongst the population and in terms of the marine and holiday amenities economy. They are densely populated and endowed with an infrastructure and productive assets and capacities that place them far above the region's inland territories.

At the same time the coastal zones with their natural estuary complexes, sea coast and shelf zones fulfil important biosphere functions (regional and global). This is where the unique natural complexes and biocoenoses are concentrated, seasonal birds migrate and sea fauna routes cross, many of which have the status of most protected objects and territories. It is precisely the sea coast, estuary areas of rivers, and the adjoining shelf habitats that are the most vulnerable ecologically.

Combining the considerable and ever increasing anthropogenic-technogenic burden on seaside areas with their specific biosphere functions and increased ecological vulnerability requires a targeted normative-legal basis governing the use of nature in seaside zones as well as stringent and well-founded regulation of the basic kinds of economic activity on the sea-coast, clear functional zoning of the sea-coast and a more rational territorial organisation of settlement and economic policy for the region.

The biosphere potential of the Ukrainian Black Sea Coast needs in-depth study and cadastral assessment. The largest national Ukrainian parks, many unique natural entities, and hugely important marshlands are situated here. The region has tremendous natural recreational potential. To preserve and protect its unique landscapes a modern normative-legal base is required, one that meshes with both national and international legislation.

In Ukraine's present state of development as a sovereign state, its Black Sea coast is but one part of the overall seaside zone, and it is necessary to solve the questions of conservation of the architectural history and the cultural and natural heritage of the region within the overarching framework of the European cultural heritage. The main aim of regional policy in this respect is to elaborate framing principles for the possible generation of a co-evolutional architectural history in the zones bordering the Black Sea. The main focus here is as delineated in significant pronouncements contained in the Rome, Athens and Venice charter, Club of Rome documents, recommendations by ICOMOS, UNESCO, and European principles of regional policy. The task involves organising architectural history and the natural-ethnic environment through the prism of conservation, recreation, and scientifically sound rehabilitation of items of architectural history and cultural and natural heritage, taking as the point of departure man as a personality together with his history, culture, architecture, and the traditional singularities of his ethnic and natural living surroundings.

3.3 Conceptual Framework of the Regional GIS

At the Department of Economic and Social Geography at Odessa State University, work has begun on elaborating a multipurpose geographical information system for the Ukrainian Black Sea Coast (UBSCgis project). The juridical basis for research is a special ruling of the Ukrainian Cabinet of Ministers on the State Programme for the Socio-economic Development of the Ukrainian Black Sea Coast (5 May 1995).

The general goal of the UBSCgis is to provide information for complex integrated planning and monitoring of spatial economic and social objects (in the context of administrative territorial units, settlement systems, intersectoral complexes and production systems) on the basis of automation of integral expert assessment in a GIS environment. In the long run, a GIS regional expert must be created which will aid decision-making in regional planning and management. The ability to work within a territorial context and provide various types of information on the state of the regional economic complex and the geographical environment in the form of generalised integral indices is the single most important factor leading the way towards enabling the principle of integrated regional planning (up to now only posited) to be turned to effect. Besides, GIS will equip the socio-economic monitoring of local and regional territorial complexes with the capabilities of forecast and management. The complex integrated geographical assessment will then be based on the results of integrative expert estimates and thematic mapping of:

- natural resources;
- population and its dynamics;
- level and dynamics of socio-economic development of administrative territorial entities;
- investment fitness and efficiency of economic entities and complexes;
- socio-economic conditions for creating free trade zones;
- level and dynamics of anthropogenic burden on the territory;
- dynamics of environmental quality;
- architectural history of northern Black Sea coast cities.

More specialised tasks of the UBSCgis project are the elaboration and implementation of an effective information infrastructure (within the regional public administration system) for monitoring and regulation of ethnosocial and ethnopolitical processes along the Ukrainian Black Sea Coast to avoid ethnic conflicts and resistance as well as to promote sustainable development in the multinational, polyethnic society of the new Ukrainian planning region. This legal objective is related closely with the object of research, namely to study the geodemographic, socio-economic, ethnocultural and ethnopolitical patterns and factors which determine the ethnosocial and ethnopolitical situation in the region. Research will need to follow two main routes:

a) geodemographic exploration of all administrative territorial units, settlement systems and major cities in the region. Available official census data will be used to design a GIS database for generation of an electronic geodemographic atlas of the region. Additional ethnodemographic field research needs to be carried out wherever there are concentrations of national minorities and ethnic groups as in Odessa, Nikolayev, and Kherson oblasts and the Autonomous Republic of Crimea. Especial attention will be paid to the study of the ethnodemographic situation in rayons where programmes for the relocation of displaced nations (Germans, Tartars) and refugees from NIS countries are underway. The electronic geodemographic atlas of the region has to be designed using GIS technology. It will provide automated thematic mapping of the geodemographic situation and its current trends as well as the requisite detailed geodemographic patterns of all national minorities and ethnic groups residing in the region;

b) ethnogeographical study of the region with the aim of clarifying ethnocultural patterns and ethnological processes in an historical-geographical context. Expeditions and field research are required at sites of concentrated settlements of national minorities and ethnic groups, i.e. Bulgarians, Moldavians and Gagausians in Odessa oblast, Germans in Nikolayev and Odessa oblasts, Tartars in the Crimea. Special ethnogeographic field research is needed to explore current conditions in the places to which displaced Germans (Odessa oblast) and Tartars (the Crimea) are being relocated. In the course of research, ethnic patterns in rural households have to be identified and various ethnodemographic, ethnocultural, ethno-socio-economic and ethnopolitical 'portraits' of all ethnic groups residing in the region obtained. This is a necessary prerequisite for ethnodemographic policy-making by the regional state administrations. Once research has been concluded, an electronic ethnogeographical atlas of the region will be elaborated using GIS and multimedia technology. It will be able to produce composite ethnographic maps accompanied by various video and audio recordings.

With this method, the data treated by the system are presented conceptually in a kind of informational 'cube' whose axes represent:

1. the requested spatial scale of socio-economic activity (the sequence 'settlement-rayon-oblast-region' for example);
2. the requested degree of data aggregation in estimations (e.g. 'raw census data-intermediate coefficients-integrated indices');
3. the requested scale of time presentation (e.g. 'current, middle, long-term' presentations or stages of the State Programme itself).

This indicates that the list of urgent and key issues of regional development in the context of regional planning and ICM is somewhat lengthy along the Ukrainian Black Sea Coast. Table 1 illustrates the dimensions of a GIS adequate to the planning and management activities in the region.

In the first stage of the UBSCgis-project the main organisational issues were explored and worked out. They include:

1. inventorisation and assessment of all possible GIS-relevant data available as well as their bearing on every aspect of regional planning and ICM issues set out in Table 1;
2. definition of priorities and interfaces of planned GIS activities;
3. selection of hard- and software platforms for GIS implementation;
4. definition of potential GIS end-users in the region and problems relating to GIS data dissemination;
5. elaboration of GIS-based network for monitoring and correction of planning and management initiatives in the region.

Table 1. Correspondences between regional planning and management issues and GIS activities

Regional planning and management issues	GIS activities
1. Elaboration of conceptions and programmes on regional socio-economic development	1. Formation of regional socio-economic GIS for needs of state management and local self-management
2. Determining primary thrust of restructuring of regional economic complexes	2. GIS-based planning survey of economic complexes and production systems
3. Elaboration and implementation of market mechanisms and structures	3. GIS-based assessment of investment fitness and commercial efficiency of economic structures and complexes
4. Generation of an inventory and cadastral register of natural conditions and resources	4. Elaboration of regional cadastral GIS, GIS-based expert assessment of natural resources
5. Elaboration of geographical cadastre on natural conditions, natural resources and quality of environment	5. GIS-based monitoring of natural conditions, natural resources and environment quality in spatial-temporal mode
6. Installation of more updated resource- and energy-saving technologies	6. GIS-based inventory and assessment of resource- and energy users in environmental context
7. Implementation of regional ethnosocial and demographic policy	7. GIS-based geodemographic monitoring, electronic ethnographic atlas of the region
8. Development of modern service and social infrastructure	8. GIS-based geomarketing, retail analysis and service location modelling
9. Analysis of contemporary geopolitical situation, elaboration of principles and methods of regional foreign economic and innovative activities	9. Small-scale automated thematic mapping, GIS-based survey of geographical and socio-economic conditions for establishing free trade zones in the region
10. Elaboration and realisation of regional nature conservation and relevant socio-ecological policy	10. Environmental monitoring using remote sensing and GIS-technology, elaboration of electronic atlas of nature reserves and protected areas
11. Harmonisation of architectural history assets of northern Black Sea coast cities and urban agglomerations	11. Designing of municipal GISs for coast cities and urban agglomerations, elaboration of GIS-based architectural history atlas of the region

3.4 Current Standing and Future Prospects

3.4.1 Geographical Database of the Region

A geographical database of the region comprising basic electronic maps has been designed on an IBM PC platform using Atlas*GIS software (Strategic Mapping, Inc.) and includes:

1. all levels of administration and territorial units of the region: local councils, administrative rayons, oblasts and the Autonomous Republic (Figure 2);
2. settlement system of Ukrainian Black and Azov Seas Coastal Zone (Figure 3);
3. large-scale map of Odessa urban agglomeration;
4. inventory map of nature reserves and protected areas of the region.

As a local-level ICM pilot project is to be started in the region (Figure 3), large-scale electronic maps of the study area are now being compiled.

3.4.2 GIS-based Economic Assessment of Natural Resource Potential

A method of economic assessment of natural resource potential by territorial units has been developed using the GIS approach. It permits computation of special integral indices with the aid of which it is possible to estimate the economic value of the natural resource potential of a given territorial unit (in sectoral or integral mode). A characteristic feature of this GIS-based method is a synergetic mechanism of quantitative assessment using geographical (locational) and economic (complementarity of related resource values) information (Figure 4).

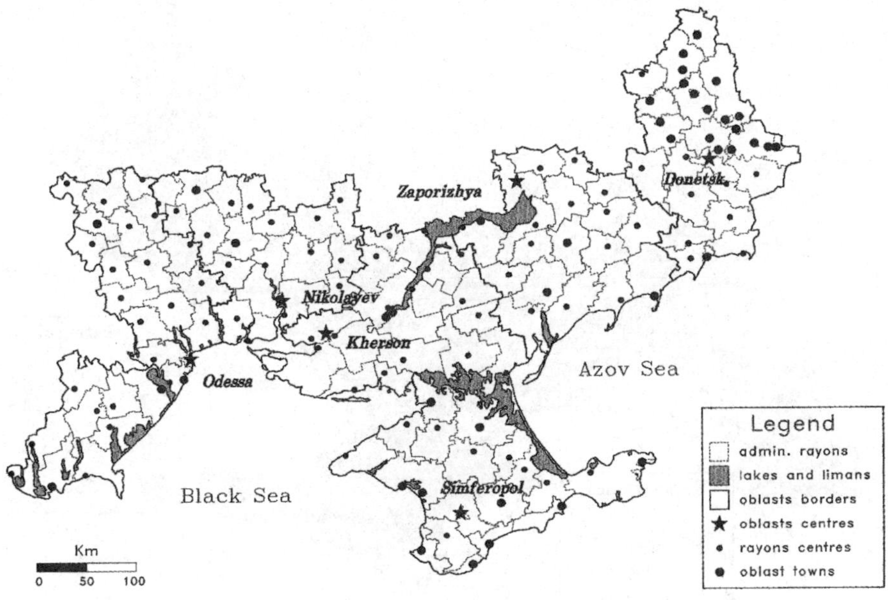

Figure 2. Ukrainian coastal area (geographic database of UBSRgis)

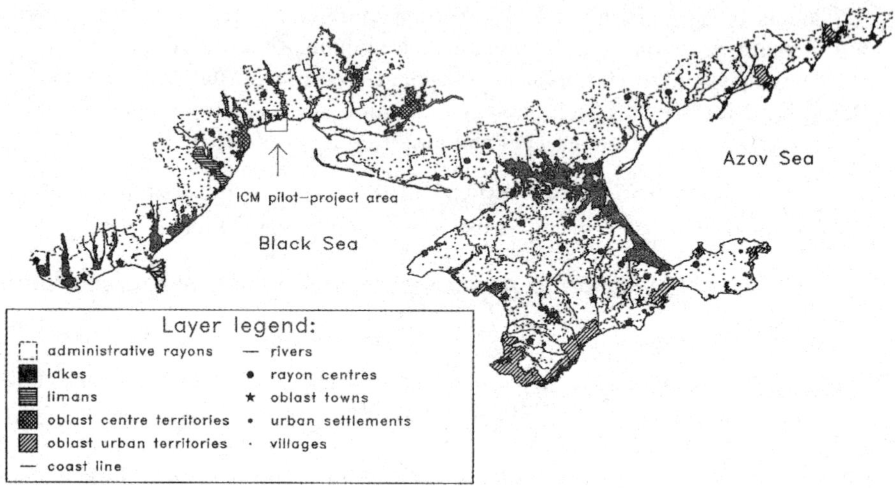

Figure 3. Ukrainian sea coastal settlement system

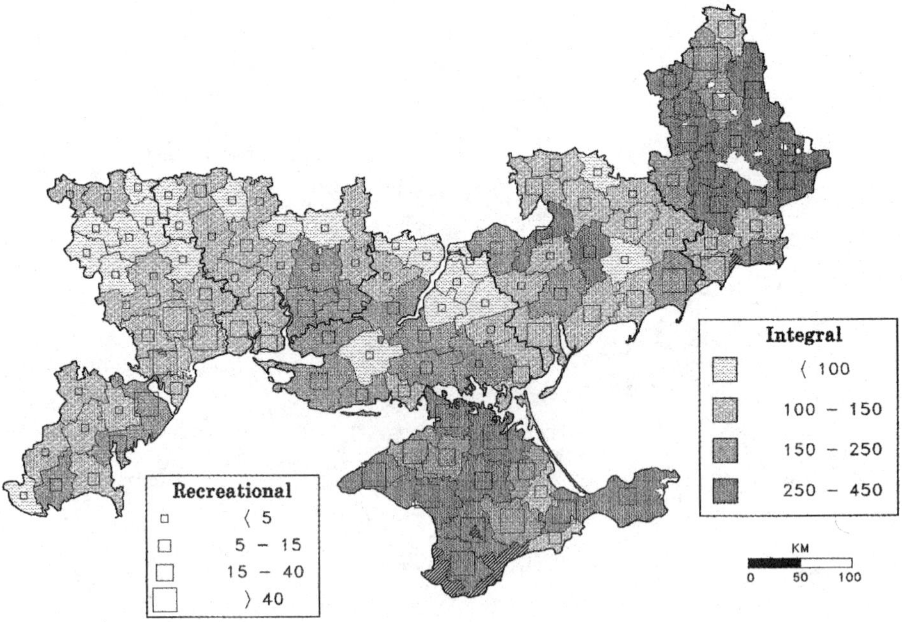

Figure 4. Economic assessment of natural resource potential (in points)

3.4.3 Geodemographic and Ethnographic Electronic Atlases of the Region

A geodemographic atlas is in preparation for spatial-temporal monitoring and analysis of demographic processes in the region. There are many specific regional factors determining demographic trends in that part of Ukraine, viz. ethnic, socio-economic, political and environmental (Figure 5). An electronic atlas with statistical analysis and modelling packages (CSS, SPSS, FLOWMAP – with specialist software developed in Utrecht university) is in use, so users can perform a set of analytical procedures, among them are:

1. analytical visualisation of raw census data by territorial units;
2. multidimensional scaling, cluster and factor analysis;
3. surface modelling;
4. migratory flux analysis;
5. spatial response modelling;
6. forecast modelling;
7. expert system approach.

The electronic ethnographic atlas of the region will be generated using GIS and multimedia technology. It will be able to produce composite ethnogeographical maps accompanied by various video and audio recordings.

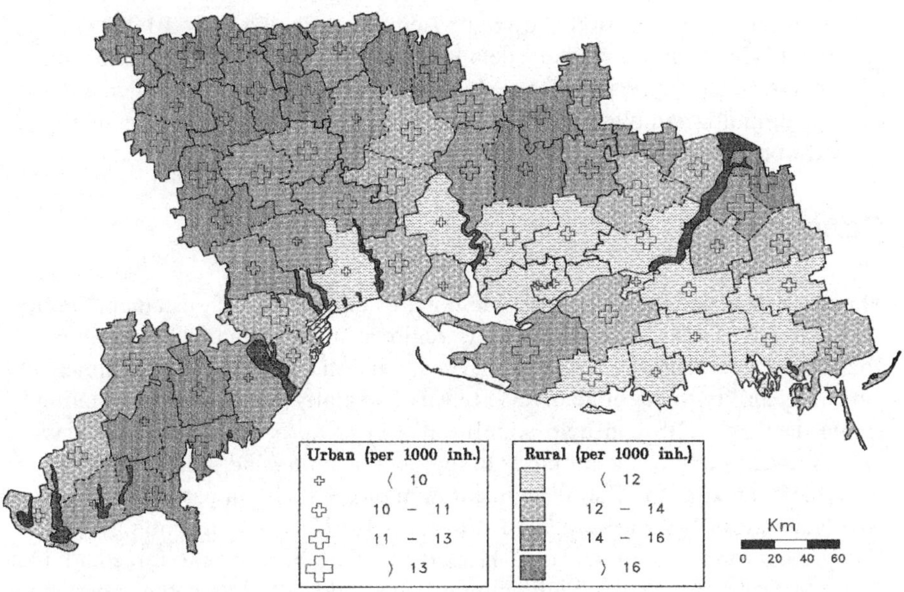

Figure 5. Population death rate in UBSR (average per year in 1991–1995)

3.4.4 GIS-based Assessment of the Anthrogenic Burden on Coastal Territories

An environmental cadastre of coastal territories (administrative rayons) has been devised using Atlas*GIS and IDRISI software. The key economic and social variables associated with and affecting environmental quality have been collated to compute integral assessment indices for all coastal administrative rayons by thematic section (agriculture, industry, infrastructure, population). Final results have been presented in the form of a set of thematic maps on the risk of potential environment pollution.

3.4.5 Medical Electronic Atlas of Nikolayev Oblast

A medical electronic atlas of Nikolayev oblast was designed using Atlas*GIS. It currently comprises more than 100 thematic maps created on the strength of public health care statistics. The main reason why Nikolayev oblast has been selected as a study case was the need to identify the possible health effects of the nuclear power station located north-west of Nikolayev and explore the geographical distribution of cholera, of which there are still cases in some coastal areas of the oblast.

3.4.6 Digital Map of Odessa Agglomeration Coastal Districts

To elaborate a GIS-based architectural history cadastre for coastal districts of the Odessa urban agglomeration, a digital basemap originally of 1 : 10 000 scale was generated. This cadastre contains detailed cartographic and textual descriptions of architectural items to be preserved or reconstructed. By combining thematic urban maps with multi- and hypermedia technologies it is possible to create an effective information-rich working environment for urban planners and architects.

Conclusion

This paper has sought to demonstrate how, in the context of sustainable regional development, GIS can contribute to regional/spatial planning and integrated coastal management of varying spatial scale and in different sectoral dimensions. Due to its ability to integrate diverse types of spatially distributed information in a single database, GIS can help significantly in the process of complex environmental assessment for the purposes of regional planning and policy making.

In the context of the European integration process and, in particular, the organised Network of Spatial Research Institutes (CEE Net), GIS technology, given its ‘integrating’ and ‘communicating’ properties, could be a driving force and a conduit for trans-European coordination in the field of sustainable development planning and policy making.

Despite widespread recognition of the benefits of GIS in regional/spatial planning, the adoption of GIS by planners has been slow to take off due to the complexity of proprietary software and an associated lack of the in-house expertise needed to use the systems now available. Therefore, appropriate GIS education is necessary to train planners in fundamental computing skills as well as in the use of the most popular GIS software. To our mind, it would be worthwhile harnessing GIS-based planning experience gained in CEE to date by organising a kind of CEE SubNet for the dissemination of GIS-based planning teachware in central and eastern Europe.

References

Anderson VN et al. (1994) Integrated environment assessment in urban and regional planning in Ukraine: some GIS applications. In: EGIS/MARI'94 Conference Proceedings, Utrecht/Amsterdam: EGIS Foundation, pp 1949–1950

Anderson VN, Topchiev AG (1993) Odessa-Utrecht GIS initiative: the human geography projects. In: EGIS'93 Conference Proceedings, Utrecht/Amsterdam: EGIS Foundation, pp 1311–1315

Anderson VN, Smolskyi VL (1994) GIS-based geodemographic monitoring in Ukraine: towards the demands of transforming economy. In: EGIS/MARI'94 Conference Proceedings, Utrecht/Amsterdam: EGIS Foundation, pp 1951–1952

Anderson VN, Polosa AI, Smolskyi VL (1993) Analytical toolbox of a geodemographic information system: South Ukraine application. In: EGIS'93 Conference Proceedings, Utrecht/Amsterdam: EGIS Foundation, pp 1516–1517

Anderson VN, Skrizhevskaya EV (1996) Solving medical geography problems with GIS: the case of South Ukraine. In: Geographical Information. From Research to Application through Cooperation. Volume 2. Edited by Rumor M, McMillan R and Ottens HFL, Publisher: IOS Press, pp 857–861

Kiib H, Veirum NE (1993) Hypermaps in urban planning. User requirements and design guidelines for HYPERMAPS in urban design and planning control in decentralised planning systems. ISSN 0902-8056 Publication number 109. EGIS'93 Conference, pp 1–11

Ottens HFL (1990) The application of geographical information systems in urban and regional planning. In: Scholten HJ, Stillwell JCH (eds), Geographical Information Systems for Urban and Regional Planning, Kluwer Academic Publishers, Dordrecht, pp 15–22

Pistun MD (1996) Basics of the theory of human geography: Textbook, Kiev, Vishcha shkola, 231 pp

Stillwell JCH, Scholten (1990) Editorial introduction. In: Scholten HJ, Stillwell JCH (eds), Geographical Information Systems for Urban and Regional Planning, Kluwer Academic Publisher. Printed in the Netherlands, pp XXIII–XXV

Integrated Coastal Zone Management around the Black Sea

Igor A. Kharitonov
International Activity Centre on Integrated Coastal Zone Management,
Global Environmental Facility (GEF),
Black Sea Environmental Programme (BSEP)
Krasnodar, Russia

Sustainable Development and Integrated Coastal Zone Management

"Sustainable development" is used to mean: improving the quality of human life within the carrying capacity of supporting ecosystems (Wager, 1995).

A "sustainable economy" is the product of sustainable development. It maintains its natural resource base. It can continue to develop by adapting and through improvements in knowledge, organisation, technical efficiency and wisdom.

According to J. F. Wager (1995), "sustainable society" observes nine principles of sustainability:

- to respect and care for the community of life;
- to improve the quality of human life;
- to conserve the Earth's vitality and diversity;
- to minimise the depletion of non-renewable resources;
- to keep within the Earth's carrying capacity;
- to change personal attitudes and practices;
- to enable communities to care for their own environments;
- to provide a national framework for integrating development and conservation;
- to create a global alliance.

The Integrated Coastal Zone Management (ICZM) system is a system for controlling development and other human activities that affect the condition of economic resources and the quality of the environment in coastal zones (Clark, 1992).

The Statement of the International Conference on Coastal Zone Management (Noordwijk, The Netherlands, 5 November, 1993) proclaims that "ICZM has been identified as the most appropriate process with which to address current and long-term coastal management issues, including habitat loss, degradation of water quality, changes in hydrological cycles, depletion of coastal resources, and adaptation to sea level rise and other impacts of global climate change. It is also a means of identifying and anticipating future opportunities. Thus, ICZM is a major tool with which to achieve sustainable development in coastal states, including states with boundaries on inland seas".

The goals and benefits of ICZM fully coincide with the goals and benefits of any sustainable development programme. The overall goal of an ICZM-type programme is to ensure optimum sustainable use of coastal natural resources, perpetual maintenance of high levels of biodiversity, and real conservation of critical habitats (Clark, 1992).

Sustainable development of the Black Sea countries aimed at increasing the living standards of the population and preserving the unique ecosystems of the region is impossible without the establishment of an effective Integrated Coastal Zone Management (ICZM) system in each country. The typical socio-economic and ecological problems and transboundary interaction within the natural and social fields necessitate close coordination of the efforts of all countries and improvement of national ICZM systems. It may be useful to establish an international ICZM body for the Black Sea countries given the importance and complexity of some countries' problems.

ICZM has global experience going back more than 20 years. The USA has the most developed ICZM programme. J. R. Clark (1992) has formulated guiding principles for such a programme:

- the coastal area is a unique resource system which requires special management and planning approaches;
- water is the major integrating force in coastal resource systems;
- it is essential that land and sea uses be planned and managed in combination;
- the edge of the sea is the focal point of coastal management programmes;
- coastal management boundaries should be issue-based and adaptive;
- a major emphasis of coastal resources management is to conserve common property resources;
- prevention of damage from natural hazards and conservation of natural resources should be combined in an ICZM programme;
- all levels of government within a country must be involved in coastal management and planning;
- the nature-synchronous approach to development is especially appropriate for the coast;
- special forms of economic and social benefit evaluation and public participation are used in coastal management programmes;
- conservation for sustainable use is a major goal of coastal resource management;
- multiple-use management is appropriate for most coastal resource systems;
- multiple-sector involvement is essential to sustainable use of coastal resources;
- traditional resource management should be respected;
- the environmental impact assessment approach is essential to effective coastal management.

As a state programme, ICZM is new to the Black Sea countries. Work in this field started in Bulgaria in 1991 within the framework of the GEF Black Sea Environmental Programme and with support from the World Bank. By 1995, all Black Sea countries had carried out a special investigation of the present national ICZM, evaluated the advantages of the existing management system and highlighted its shortcomings. Corresponding national reports were submitted to national government bodies.

Boundaries of the Coastal Zones

All Black Sea countries keep to the definition recommended by World Bank consultants: "A coastal zone is the area where the processes which depend on the sea/land interaction are at their most intense." Such descriptive unanimity does not prevent the countries from adopting different methods of setting the extent of their coastal zones. The procedure for defining the inland boundary may be based on three different approaches: 1) scientists employ the basin approach; 2) in many countries where the ICZM programme is in action at government level, as it needs to be, the coastal zone is established on the basis of administrative boundaries; 3) the third approach involves defining coastal zones on the basis of some traditional, historical or other non-administrative boundary, for example, zone of specific economic activity, etc. A combination of all three approaches in different areas of the coastal zone of a given country is also feasible.

The boundary of the coastal zone at sea may be prescribed to run either along the boundary of the economic zone or zone of territorial waters, along the external shelf boundary, or along the normative non-administrative boundary.

At present many problems of nature use in coastal zones are caused by their complicated system of internal boundaries and management tools. There are climatic, geological, geobotanic, zoogeographical, landscape and other natural boundaries. There are also a great number of boundaries established through directives. These include the boundaries of special economic zones along the coastal line, the boundaries of nature reserves, protected areas, natural monuments, water protection zones, forestry reserves, hygiene zones for enterprises, private cottages (dachas) construction zones, health-resort zones, residential zones, territorial waters etc. Given legislative and organisational simplification of the existing intrazone boundaries, they should cease to be a source of conflicts of interest between different nature users, since management would be integrated. The main importance would be accorded to newly drawn "perpendicular" internal boundaries of different ICZM agencies and to external coastal zone boundaries.

The basic criterion in the establishment of coastal zone boundaries is the presence of a geographical space where natural and human processes of land-impact on sea and sea-impact on land are to be observed that pose problems requiring an ICZM system for their solving. Ideally, land boundaries in coastal zones are established as deep into the country as is necessary for successful attainment of

ICZM aims. Thus the boundary line can shadow the whole watershed of river basins or else reduce the coastal zone to a narrow strip along the water line if it is sufficient for ICZM problems solving.

ICZM is a governmental programme designed to coordinate the actions of diverse state institutions, functioning within existing administrative boundaries. That is why the coincidence of coastal zones boundaries with existing administrative boundaries would greatly facilitate the setting-up of an ICZM system.

The variant of coastal zone external boundary proposed for the Black Sea countries reflects territories having different nature use problems within them caused by the specifics of matter and energy transfer in ecosystems. Thus Black Sea territories with administrative boundaries along the watershed of the Main Caucasus Mountain Ridge are relatively autonomous regarding ecological regulation as they comprise whole river basins. By contrast, the Azov Sea regions of Russia and the Ukraine are only final links in a complex chain of matter and energy transfer covering large areas of the Don, Kuban and smaller steppe river basins far beyond the boundaries of the coastal zone.

A detailed regulation-of-boundaries regime taking into account all the processes of cross-boundary transfer is necessary for successful implementation of ICZM principles.

In spite of the differences between regions in the Black Sea coastal zone adumbrated, it is possible to apply a single principle of coastal zone boundary demarcation to the zone as a whole.

The coastal zone of the six Black Sea countries is an area 50 km wide (land) embracing 10 higher administrative units (province, region, oblast, autonomous republic) and elements of a further 23, these in turn being made up of 305 lower administrative units (district, town, municipality). The administrative and basin approaches were used to set land boundaries. In countries where there are mountain ranges along the sea (Turkey, Russia), the application of multiple approaches – administrative and basin - yielded similar results, since in their natural conditions administrative divisions reflect the relief of the area.

Accordingly, the Black Sea and Azov Sea coasts are under the jurisdiction of the authorities at four levels: national, regional, municipal and local. The only exception is Turkey, where a considerable part of the coastal zone is managed by provincial government directly.

The marine boundaries have been defined in all Black Sea countries, but using differing principles. The Bulgarian, Russian and Ukrainian boundaries correspond to the 12-mile zone, the Romanian to the 50-mile zone, the Georgian to the 5-mile zone, and the Turkish to the entire Turkish zone of the Black Sea.

The boundaries of the coastal zone have been legally approved only in Bulgaria and Russia.

In Turkey the entire area of greater Istanbul was excluded from the Black Sea ICZM project because of the importance of this city in the socio-economic field. This may create difficulties regarding assessment of human impact on the marine environment and the solving of various other questions since the natural and socio-economic systems of this conurbation are interconnected with nearby areas.

ICZM Issues

The ICZM process is represented in the form of the following triad: management subject – management object – management tool (accordingly: management body – coastal zone plot – legislative base). As yet, no member of this triad has been adapted to ICZM aims.

In the legislative sphere in Russia, for example, there are more than 30 normative documents and laws regulating the use of natural resources in coastal zones. Nevertheless none of these documents defines the status of coastal zones or establishes the nature of their management. Laws and normative documents in force at six levels – international, federal (presidential), governmental, regional, municipal and local – combine to create a system that is very close to chaos. In the course of organising the ICZM system, the legislative base should be significantly regulated and a special management body (national and/or regional) should be given the function of executor of a coastal zone laws complex.

The coastal zone management body would need to have a complex remit. Not only would it replace dozens of organisations of the various ministries and departments as well as regional and local management institutions that function at present in this sphere, it would also set out the goals of the ICZM. A management body should not simply represent a totality of functions or subjects. Management may be considered to be integrated only when, having successfully established an algorithm, it generates an overall performance that exceeds the sum of the input produced by the elements that go to make up the body. This is the type of synergetic effect that the transformation of the existing system of coastal zone management should be aiming to attain. Formation of the relevant management body is the most difficult task facing organisers of an ICZM system, as here the vested interests of any number of management structures and departments that have taken shape over the decades are called into question.

The system-forming element of the above mentioned triad is the management object – the coastal zone. It is the resources, problems and real situation of the latter that will define the range of problems to be submitted for integrated management on the basis of optimised legislation.

Where international seas are concerned, the ICZM system can only be implemented with recourse to international collaboration. In fact, it is only through the cooperation of all Black Sea countries – through recognised international legal, institutional and political channels – that it will be possible to obtain meaningful management and socio-economic results.

International Collaboration

The Global Environmental Facility (GEF) is called upon to play an important role as a multifaceted financial tool for solving environmental problems in countries searching for new structures of economic development. Participating in the GEF signifies that a country or a larger institutional framework has chosen to improve the efficacity of funding mechanisms for projects with a pronounced environmental dimension.

The Black Sea Environmental Programme (BSEP) was formally established in September 1993. The programme is funded by GEF with additional cost sharing contributions from CECs and the PHARE and TACIS programmes as well as bilateral contributions from Canada, the Netherlands, Switzerland and France.

By early 1994, a Programme Coordination Unit (the PCU) had been established in Istanbul and a work plan had been agreed by its National Coordinators (in many cases Ministers of the Environment or their deputies).

According to L.D.Mee (1994), the programme has three primary objectives:

- to strengthen and create regional capacities for managing the Black Sea ecosystem;
- to develop and implement an appropriate policy and legal framework for the assessment, control and prevention of pollution and the maintenance and enhancement of biodiversity;
- to facilitate the preparation of sound environmental investments.

Each of these objectives has enormous scope and involves the participation of a broad range of sectors. The BSEP Steering Committee incorporates the National Coordinators, donor representatives and representatives of NGOs (including the business sector). It is clear that a mechanism is needed to share the burden of these programme objectives if the Black Sea Environmental Programme is to lead to a long-term Action Plan. Rather than expect the coastal countries to contribute to a single centralised future programme, it was decided to instigate a series of Working Parties based upon "Activity Centres", which are national institutions that already have the basic infrastructure and personnel to coordinate specific tasks in the region. Each Government has agreed to host one of these centres.

The Activity Centres and their Working Parties are as follows:

1. Emergency Response (Varna, Bulgaria);
2. Routine Pollution Monitoring - i.e. the assessment of land-based sources, bathing water quality, etc. (Istanbul, Turkey);
3. Special Monitoring Programmes, Biological and Human Health Effects and Environmental Quality Standards (Odessa, Ukraine);
4. Protection of Biodiversity (Batumi, Georgia);
5. Development of Common Methodologies for Integrated Coastal Zone Management (Krasnodar, Russia);

6. Fisheries (Constanza, Romania).

In addition there are three Working Parties based at the PCU:

- Data Management and Geographic Information Systems (GIS);
- Advisory Panel on the Harmonisation of Environmental Quality Criteria, Standards, Legislation and Enforcement;
- Environmental Economics Expert Group.

The ICZM Activity Centre in Krasnodar was founded in 1993 to implement the ICZM component of BSEP. The mandate of the Centre, as defined in the BSEP Project Document, is "to facilitate the exchange of information and experience on ensuring sustainable resource use, including recreational use by tourists in the coastal zone of Black Sea countries and to develop methodologies for coastal zone management, with particular reference to threats to the environment arising from the transition to market economies".

The main tasks of the Centres are identified as:

- assisting Black Sea countries in formulating and implementing National ICZM strategies, policies and plans;
- assisting authorities, institutions and experts involved in ICZM in developing and implementing appropriate ICZM methodologies, tools and techniques;
- coordinating and harmonising activities and comparing notes with other BSEP Activity Centres;
- cooperation with UN agencies and other organisations involved in ICZM.

ICZM Activity Centres are coordinating the activity of the National Focal Points of Bulgaria, Georgia, Romania, Russia, Turkey and the Ukraine for the establishment of boundaries to coastal zones, the reform of national legislation and the institutional base, and tackling the main ecological and socio-economic problems.

Status of the BSCZ and Priority Actions

In 1996, the first drafts of the National ICZM Assessment Reports were submitted to the World Bank for review and comment. The final drafts were sent to the ICZM Activity Centre, which prepared a draft of the Regional ICZM Synthesis Report. The national ICZM reports collated information on the state of the given country's natural environment, socio-economic and demographic parameters, the status of national and local policies and legislation, and issues pertaining to land and marine use impacts on the coastal zone. The reports also included the local definition of geographic and administrative boundaries of the coastal zone and identified the importance of each government establishing official land- and marine-based boundaries. Reports showed that there were several common issues with respect to the coastal zone among the Black Sea countries:

- in all the Black Sea countries, the historical/archeological heritage was significant and substantially threatened;
- all Black Sea countries were experiencing economic difficulties;
- the economic problems have in turn generated social problems (i.e. unemployment);
- there were land ownership and land management problems in the countries in transition to market economy.

Different types of tourism development within the coastal zone had been initiated (international tourism in Bulgaria and Romania, national tourism in Russia and Ukraine, local tourism in Turkey). The state of the coastal zone environment was not considered to be entirely healthy, but not catastrophic either. Natural resources were being subjected to significant degradation and exploitation from economic activities. Complete geo-information systems were in place in only three out of six countries.

Special laws on ICZM are being developed in many countries with the anticipation of being adopted as legislation. Bulgaria was successful in developing ICZM legislation and submitting it to parliament, where it is currently being considered. An institutional framework for coastal zone management was identified as a necessary element for successful integrated management. Bulgaria, for instance, has initiated a framework which creates a special ICZM structure within the national and local government system.

These reports were the first attempt to synthesize existing conditions and identify the multi-sectoral issues that affect the coastal region within each country. As substantial background documents, the ICZM national reports also represented a contribution to the Black Sea Environmental Priority Studies.

Drawing on the national reports, the ICZM Draft Synthesis Report, prepared by the ICZM Activity Centre in Krasnodar, Russian Federation, was presented and approved during the third ICZM Working Party. The report was revised and published as the "Summary Report on Black Sea Integrated Coastal Zone Management". This report was also used in the preparation of the Black Sea Transboundary Diagnostic Analysis (TDA). Overall, the synthesis report stressed the importance of the continuing development and implementation of ICZM methodology and its practical application both in individual countries and in the region, the importance of developing national ICZM strategies, and more efficient administrative bodies in terms of training, management and administration. Geo-information equipment training was recommended during the second phase of BSEP.

The present state of the Black Sea coastal zone is identified in the Regional ICZM Synthesis Report, as are priority actions.

1. Environmental Issues:

- anthropogenic pollution;
- disposal of municipal waste;

- limited number of protected zones;
- exhausted fish stocks;
- coastal erosion.

2. Legislation and Institutional Issues:

- legislative and institutional weakness of ICZM;
- need for updating national legislation;
- absence of Geographic Information System (GIS);
- poor land management practices and weak institutions.

3. Socio-economic Issues:

- depressed economic conditions;
- structural problems in industry;
- unemployment;
- outdated equipment, primarily in ports;
- lack of investment.

Priority Actions accordingly:

1. Environment:

- to update water supply and water treatment;
- to upgrade waste disposal and reduce the dumping of solid and marine wastes;
- the construction of sewage treatment plants;
- the promotion of industry, transport and agricultural services;
- the creation of a system of protected sea areas.

2. Legislation and Institutional Base:

- to form a legislative base for the Black Sea ICZM system;
- to strengthen national and local institutional capacity;
- to finalise the legislative base for property issues;
- to reconstruct real coastal management systems;
- to adapt national legislation to ecological needs;
- to adopt a set of normative acts on economic regulations for natural resources and set up protected areas within the coastal zone;
- development and approval of the Black Sea Strategic Action Plan (BS-SAP).

3. Socio-economic Issues:

- to set in train an economically self-sustainable system of land and nature use;
- to develop a sustainable tourism strategy to attract investment in tourism and recreational amenities;
- to promote the efficient use of transportation;
- better agricultural and forestry management;

– to develop agriculture.

The Strategic Action Plan for the Black Sea (BS-SAP) was adopted at the Ministerial Conference held in Istanbul, Turkey, 30–31 October, 1996. The approval of the BS–SAP reinforced the active participation of the coastal countries in ICZM. In addition to the development of national ICZM strategies, a regional ICZM strategy will also be developed for the Black Sea. This regional strategy will further elaborate basic principles and methodologies for land and water use planning as well as zoning systems. The methodologies and principles recommended in this regional strategy will be taken into account when developing or reviewing national strategies and planning for ICZM. The BS–SAP also establishes intersectoral committees for the ICZM at national, regional and local levels of public administration as well as the Advisory Group on the Development of Common Methodologies for Integrated Coastal Zone Management, coordinated by the Activity Centre in Krasnodar, Russian Federation.

References

Clark JR (1992) Integrated Management of Coastal Zones. FAO Fisheries Technical Paper. vol 327, Rome

Mee LD (1994) The Black Sea Environmental Programme. Saving the Black Sea. Issue 1. Sept., Istanbul

Wager JF (1995) Sustainable Development in the Use and Management of Coastal Resources. Materials of the 1st Training Workshop on ICZM, EIA and Audit, Krasnodar

Socio-economic Consequences of the Chernobyl[1] Catastrophe and Regional Development in Ukraine[2]

Irina Gukalova
Institute of Geography, National Academy of Sciences of Ukraine
Kyiv (Kiev), Ukraine

At present, environmental issues are given insufficient attention by society and occupy lower stages in the hierarchy of values than economic issues. Few people are cognisant of the fact that humankind, as any other variety of species, has its own ecological niche, i.e. its own system of relations with the physical environment, the laws of which should be considered in planning human activities.

The term "sustainable development" which began to be widely used after the UN Conference on Environment and Development in 1992 in Rio de Janeiro, is understood as a concept of transition to such a state of nature and society which may be called "co-evolution". Today there is no other way for humankind and individual nations and social groups to survive than by adopting the route of sustainable development. Unfortunately, there is no consensus opinion as to how such development should be facilitated in different countries, there is no clearly unified definition of the subject. At every stage of its development, a society may change considerably, and its sustainability or unsustainability with it. The only thing which remains unchanged is concern over preserving the natural and cultural heritage for coming generations.

At the contemporary stage of development it is assumed that there can be no purely technological solution to the problem. More concrete changes are necessary in the overall attitude to the issue of development. It is important to recognise the fact that most natural resources and conditions for development on earth are

---

[1] 'Chernobyl' is the transliteration of the name of this Ukrainian city from Russian. It is used in the title of the contribution because it is the most commonly acknowledged form. Nonetheless, in the body of the text names of cities, regions, and rivers in the Ukraine are transcribed from Ukrainian (e.g. Chornobyl as opposed to Chernobyl, Kyiv as opposed to Kiev).

[2] The present article is based on research conducted at the Institute of Geography in the National Academy of Sciences of the Ukraine and submitted for a Candidate's Degree. The object of research was the socio-ecological situation as a combination of processes emerging in society after the sharp decline in the environmental situation. The topic of research was 'Theory and methodology of an analysis of the socio-ecological situation in the zone of the Chornobyl disaster, and corroboration of its findings'.

exhausted and human knowledge remains one of the few growth areas in the drive to perpetuate human existence.

All the above relates to Ukraine, one of the largest national territories in Europe, which has significant natural resources and unique natural conditions but is characterised by an extremely unfavourable environmental situation. Ukraine and its individual regions may be viewed as having no match in the world as regards the combination of such a vast land mass and such a vast number of environmental problems. One of the latter involves the consequences of Chornobyl - the world's worst-ever technogenic catastrophe, which occurred on April 26, 1986.

A decade has passed since the event occurred, sufficient time conceivably to have overcome the "Chornobyl syndrome" but not the social consequences. Moreover, according to many scientists, the worst is still ahead – the long-term effects of the catastrophe on health are, together with other negative developments, already making themselves felt.

The uniqueness of the situation following the Chornobyl catastrophe relates not only to the vast area over which nuclear contamination was spread but also to social, economic and political processes resulting from the event. Chornobyl became not only a technogenic catastrophe for Ukraine but also had grave social and economic consequences, as unfavourable environmental factors began to undermine many socio-economic developments. The rehabilitation of contaminated regions is taking place under conditions of a certain finiteness of territorial, natural and labour resources.

1 Effect of the Chornobyl Catastrophe on Industrial Activity

The territories affected by Chornobyl have to be viewed as a spatial expression of various stages of natural and anthropogenic degradation. Together, they compose a region with specific peculiarities in manufacturing activity. In stark contrast to regions where environmentally catastrophic situations have been accumulating over a long period of time (subsequent to the formation of population centres, major cities or intensive rural development areas), the Chornobyl zone of nuclear contamination was created virtually in an instant.

Ukraine received an overall dose of around 300,000 curies of radioactive caesium, of which 270,000 curies was disseminated over a 20 % area of the country (more than 120,000 sq.km). Over 7 % of Ukrainian land mass, the density of contamination exceeds 1 Ci/sq.km. After the catastrophe, moreover, there were 800 temporary nuclear waste sites (mainly in the Kiev and Zhytomyr regions) in addition to the wreckage of a reactor containing nuclear fuel (National Report of Ukraine 1996).

Uneven patterns of radionuclide dissemination from the destroyed reactor, the complicated trajectory of contaminated air flows, differences in landscape and climatic conditions combined to produce effects complex in configuration, effect and radionuclide content. Figure 1 shows the main zones of Caesium-137 contamination in Ukraine. The most contaminated regions are Kyiv, Zhytomyr, Rivne, Volyn, Chernihyv and Vinnitsya.

A case apart is the "exclusion zone" – the area that suffered the greatest contamination and was completely evacuated and de-industrialised.

The Chornobyl catastrophe seriously depleted Ukraine's stock of natural resources. Analysing the breakdown of Ukrainian natural resources as a whole (Marinich 1990), the most considerable losses result from contamination of land, water, and natural recreational amenities. There are also considerable losses with respect to building resources. Such circumstances pose a serious threat to moves to optimise land use and hence tap the potential of the regions.

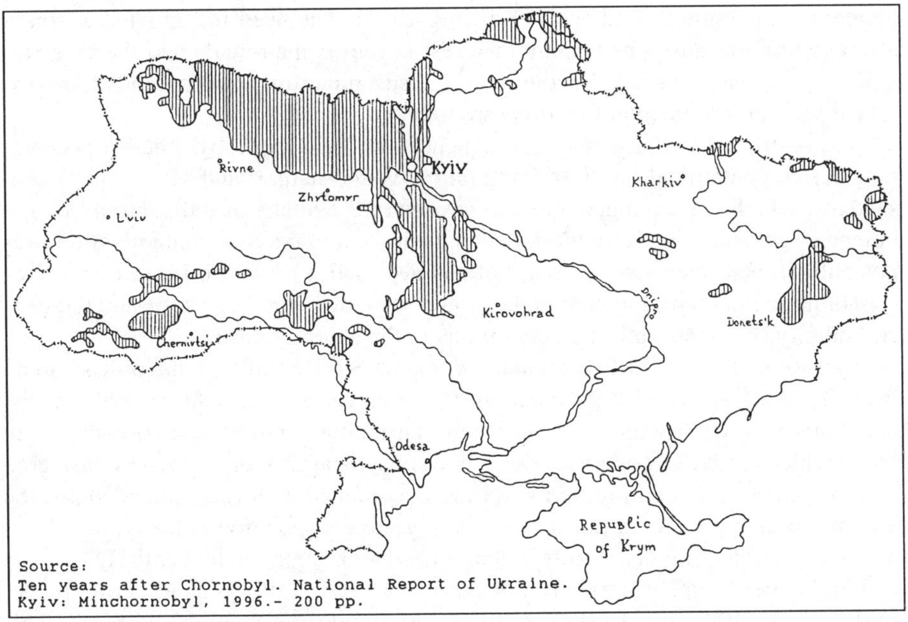

Figure 1. Contamination of Ukraine by Caesium-137 (more than 0.54 Ci/sq.km)

Within the "exclusion zone" almost all enterprises were closed down, except for the nuclear power station (Block 4) and a few other industrial sites and factories.

The Chornobyl catastrophe led to a considerable decrease in electricity production. Changes in the regional structuring of electricity generation have occurred as a result of destruction of the Chornobyl utility, followed by a moratorium on nuclear power stations imposed by the government. The Chornobyl catastrophe significantly dented public confidence in the safety of nuclear power stations. It has had repercussions on socio-economic decision-making, and yet nuclear stations now meet more than 35 % of Ukraine's energy needs (Statistical Yearbook of Ukraine 1996). The proportion of electricity generated at nuclear power stations is constantly growing. The concentration of such stations in the Dnieper river basin and other important areas is proof of an inability to construct new stations there and expand the existing ones.

Considerable damage was inflicted on woodlands by radioactive contamination. This has produced several scientific, environmental, and social tasks for the forestry industry. Among them are: evaluation of forests as a possible means of protecting areas from nuclear contamination; elaboration of a system of forestry development under conditions of nuclear contamination; protection of workers from radiation exposure; combating fires in contaminated areas etc. Such measures, which are of central importance to the forestry industry, cannot be fully implemented in contaminated areas (Kaletnik 1991). The need for targeted action is also essential because one hectare of forest in two is man-made and therefore requires special supervision. Encouraging the sustainability of woodlands is necessary if their environmental functions are to be preserved.

One of the most dangerous consequences of the Chornobyl catastrophe was radioactive contamination of agricultural lands and natural and semi-natural ecosystems, which was conducive to an increase in the intake of radioactivity by the human organism. The agricultural dimension of radioactive contamination was not only spread over a wide area, but has also had a profound impact on socio-economic factors, employment patterns and, by extension, life styles and the general quality of life as well as on conditions in the natural world.

Chornobyl led to Cs-137 contamination of 8.4016 million ha of Ukrainian land, 21 % of all agricultural lands in the country. Such conditions call for the elaboration of special measures to enable agricultural production to continue in areas which suffered contamination, and such measures have to be considered with respect to private farming as well. In accordance with recommendations, the volume of arable lands in the contaminated zone dropped dramatically, mainly as a result of production in severely contaminated areas being halted entirely.

Land under tillage decreased in extension by 1.648 million ha (5 %) in the period between 1985 and 1994. Significant structural modifications occurred. This largely relates to a reduction in land given over to food and industrial crops and an increase in land used for forage crops. These trends correspond to general recommendations issued for agricultural activities in contaminated areas. At the same time, this means cutting back more productive plant-growing activities in regions where they previously determined the status of the region as regards the geographic distribution of the labour force. Accordingly, considerable changes have been observed in the plant-growing trade.

There has also been a considerable reduction in productive cattle in the contaminated regions of Ukraine. The biggest changes occurred in the state sector, although they also affected the development of private cattle-breeding. Output from major cattle-breeders has decreased.

One of the main problems in cattle-breeding is control over the degree of contamination in production processes, particularly because the accumulation of radionuclides in milk and meat starts at the level 1.0–1.5 Ci/sq.km. Dairy cattle are particularly susceptible to radiation. Relatively less affected are pig and poultry as their food intake can be controlled.

It is important to mention that, in the given conditions of economic, environmental and energy crisis, food supplies from other regions into the contaminated zone have dropped significantly. However, it is impossible to ensure that the local population can sustain itself on products grown within the contaminated zone. Under conditions of intensive contamination, a concept of regional self-sufficiency in food products is not only hard to implement in terms of regional labour divisions, but is also a health risk.

2 Impact of the Chornobyl Catastrophe on Social Processes

The Chornobyl catastrophe, though essentially an environmental factor, has also had a decisive impact on the development of many social processes. Social change first manifested itself in the contaminated areas, then spread over the whole of Ukraine.

Forced migration of the population to less contaminated regions was the effective starting point of change. Migrations assumed the form of evacuations and the relocation of residents away from contaminated areas. In recent years, reverse migrations have begun to occur, i.e. the voluntary return of residents to their previous places of abode.

The main migration routes are perceived to run counter to the main incidences of radioactive fallout. The populations of the Kyiv, Zhytomyr and Chernihyv regions migrated to Sumy, Poltava, Kharkiv, Kirovohrad, Mykolaiv, Odessa, Kherson, Ternopil, Vinnitsya, Chernivtsi and Volyn in Ukraine. From 1985–1995, more than 142,000 persons (54,900 families) were moved from 112 residential units of compulsory resettlement. By 1 January 1996, there were still more than 5,000 families residing in such zones, including 1,000 families with children.

Forced migration has brought with it a redistribution of population within the contaminated regions. The number of people living in the Kyiv and Zhytomyr regions has decreased, though population levels there were already falling before the catastrophe at Chornobyl. Population growth was previously typical of regions to which migration then started to occur.

Urban populations were resettled in Kyiv and Chernihiv. This did not affect the structure of the urban population, although it did exacerbate the problem of housing. The makeshift town of Zelenyj Mys, which was built for workers at the Chornobyl Atomic Plant, would be unable to meet longer-term residential needs.

For this and other reasons, another town called Slavutich has been built to the north of Chernihyv *oblast*, 50 km from the Chornobyl utility. Although environmental conditions are far from ideal there and the future of the Chornobyl plant is unknown, the population of the town is growing and as of 1 January 1995 stood at 24,500. The main *raison d'être* of Slavutich derives from the "exclusion zone" and the Chornobyl plant, both situated in the Kyiv region. For this reason, Slavutich has been made an administrative unit of greater Kyiv.

The rural population from the exclusion zone has been settled within the Kyiv region. The majority of resettlers feel a sense of affinity with their new villages, which had a developed social infrastructure and which have doubled and trebled in size as a result of resettlement. Resettlement processes have had a positive impact on the rural development of villages.

Further major zones of resettlement have been established in the Kyiv and Zhytomyr regions, and also in some other regions. It is hard to determine the impact of resettlement on local population structures as the process is still underway.

It should be mentioned that in most cases the policy of resettlement did not pay much attention to the principles and priorities of the process. It generated numerous social problems, which were compounded by the psychological trauma experienced by resettlers. According to the data of a sociological survey conducted in 1992 (Figure 2), young resettlers do not give their new conditions a particularly high rating. Senior age groups provide even more pessimistic assessments. This demonstrates how difficult it is for elder citizens to adapt (Mischenko 1993).

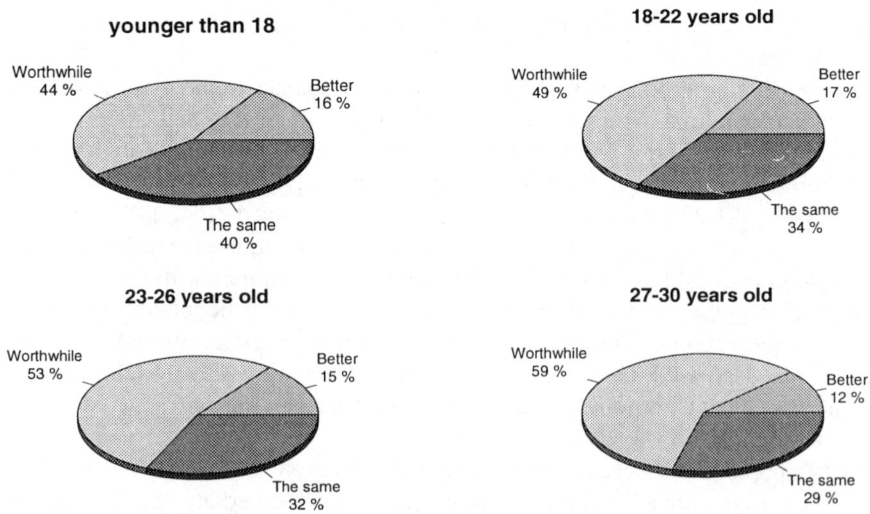

Source: Mischenko (1993).

Figure 2. How do you value your life after changing your place of residence?

Currently, the consequences of Chornobyl are no longer considered as a number one problem for the population, having been replaced by a social and economic crisis which is pushing other issues into the background (Soboleva 1996). This does not, however, mean that the catastrophe has not affected the views of people and the way they approach their current socio-psychological circumstances. The events of Chornobyl continue to be a stress factor with a long-term impact, especially since far more people consider themselves to have been affected than has been officially acknowledged.

Migration from the contaminated zones yielded a change in population structures, both at points of departure and points of arrival. For the latter, a considerable decrease was registered in numbers of women of fertile age, children, and citizens of working age, accompanied by an increase in numbers of persons of advanced age. A general ageing of the population was induced. The migrating population will not ensure adequate numbers in reproduction in places of resettlement (as numbers of senior and single citizens predominate). Therefore, ageing of the population becomes an important factor which will determine socio-economic developments in both source and target regions.

Also negative is the impact of the Chornobyl catastrophe on natural patterns of population migration. This is evidenced by an increased incidence of mortality, a decline in the birth rate in contaminated areas, and rising rates and changing structures of morbidity. The depopulation process can be said to show a tendency to increase when population decreases in monitored regions occur not only as a result of migration but also because the absolute mortality rate outstrips the birth rate. General tendencies of natural reproduction are shown in Table 1. These figures make clear that the mortality rate in all *oblasts* of the zone considerably exceeds Ukrainian mean (especially in rural areas), whilst the birth rate in the region is considerably lower than the average in Ukraine.

Table 1. Birth rates, mortality rates and rates of natural growth of population in *oblasts* of the Near-Dnieper in 1994 (per 1,000 persons)

Region	Birth rate			Mortality rate			Natural growth		
	1	2	3	1	2	3	1	2	3
UKRAINE	10.0	9.3	11.6	14.7	12.8	18.8	–4.7	–3.5	–7.2
Zhytomyr	11.8	11.9	11.7	15.8	11.5	21.1	–4.0	0.4	–9.4
Kyiv	10.0	10.2	9.6	15.8	11.0	22.2	–5.8	–0.8	–12.6
Cherkassy	10.0	10.1	9.9	16.7	11.7	22.8	–6.7	–1.6	–12.9
Chernihyv	8.9	9.8	7.6	18.7	12.9	26.4	–9.8	–3.1	–18.8
City Kyiv	8.1	8.1	–	10.8	10.8	–	–2.7	–2.7	–

1 – total, 2 – in urban settlements, 3 – in rural areas
Source: Statistical Yearbook of Ukraine (1995).

The impact of the Chornobyl catastrophe on the birth rate was especially visible in the immediate aftermath, and particularly in monitored regions. For example, in the second year after the Chornobyl catastrophe, the birth rate in the Kyiv region decreased by 13 % compared to 1986, in the Polissya rayon by a factor of 1.4, and in Ivankiv rayon by a factor of 1.3. In most of the monitored regions, there was a decrease in the number of births with simultaneous growth of induced abortions, which peaked in 1987. Thus, the ratio of abortions to births was 2.95 in the Polissya rayon. Taken together with the social-psychological situation, this parameter is considered a reflection of the unwillingness of parents to have children on account of the area's radiation history, and it impacts on the birth rate.

In Ukraine in general, and in the contaminated areas in particular, there has been an increase in the mortality rate. Mortality is increasing among the adult and child populations regarded as having been affected by the Chornobyl catastrophe (evacuees, persons engaged in decontamination work, inhabitants of contaminated territories and children born to these). It proves the fact that the radioactive factor does impact on medical and demographic processes, but that this is more clearly visible in contaminated regions or with respect to certain categories of the population affected.

From a social and psychological point of view, life in the contaminated regions is full of problems – much more so than anywhere else in Ukraine. People there have to be constantly aware of the environmental dangers which surround them and take daily precautions. For resettlers life is full of complicating factors as they switch to a new life, encountering unhelpful attitudes from people in their new places of abode and with no clear perspectives for the future as regards work etc. Such a socio-psychological set of circumstances exacts its toll on health and reproductive patterns, lifestyles, social adaptation and work activity. Time by itself is unable to cure the psychological wounds that accompany resettlement. Besides, there is little trust among the population towards current social policy in Ukraine.

Unemployment is considered one of the major problems for resettlers. Though it has become an issue throughout Ukraine, in the area affected by Chornobyl it has taken on specific features in the light of the extremely deleterious radioactive situation in the zones of contamination and the consequences of large-scale migration. Fluctuations in the labour force which were characteristic for the resettlement regions before have become more pronounced, whilst the unavailability of employment has increased as a result of forced migration.

All this speaks in favour of a more sensitive social policy towards the population affected by Chornobyl, one that is directed towards developing market resources rather than supporting the currently inefficient system of allowances and compensation schemes that can do nothing to improve the situation.

3 Specific Nature Conservation Measures after the Environmental Catastrophe

Considering the specifics of the regional environmental situation, nature conservation as a complex ensemble of scientific, technical and other measures to sustain the environment has a significant role to play in the socio-economic development of Ukraine. Nature conservation is shown to be the only way to rehabilitate the areas affected by radioactive contamination and preserve the health of their inhabitants.

Radioactive contamination of areas and entities within the national nature reserve scheme was extremely high. Intensive contamination was suffered by the Polissky (Zhytomyr region) and Kaniv (Cherkassy region) conservation areas, which cover an area of 20,104 and 2,027 hectares respectively, by the Dniprovsko-Teterivske forest and hunting reserve, as well as by a great number of further protected status areas, places of historic importance and dendroparks at local and state level.

In areas subjected to permanent radiological monitoring, conservation activities have shifted towards observation of radioactive effects on flora and fauna, monitoring numbers of species and plants, and limiting the spread of contamination to comparatively unaffected areas. This especially concerns the Zhytomyr and Kyiv Polissya regions. In other areas, nature conservation activities have proceeded as usual (i.e. without the shift referred to above), with conditions arising from higher radioactive levels always being borne in mind.

Since 1986 there has been environmental monitoring in the exclusion zone with the aim of collecting, processing, analysing, exchanging and storing information and data on levels of environmental contamination, forecasting its patterns, and elaborating scientific recommendations for decision-making in regular or extreme conditions. Specific aspects of the zone monitoring remit are as follows:

- monitoring the radio-geochemical state of land and air and of levels of concentration of radioactive substances in surface and underground waters;
- monitoring local sources of radionuclide contamination;
- overseeing subterranean or temporary radioactive waste dumps, of which there are three and 800 respectively;
- monitoring the "ukryttya" (sarcophagus) structure and its impact on the environment;
- monitoring the health of personnel and "self-settlers" and the sanitary-hygienic state of the infrastructure etc.

One of the main measures involved in tackling the consequences of the catastrophe was the localisation of radioactive materials emitted from the destroyed Chornobyl plant reactor. Within two years of the catastrophe, 21 sq. km of surface had been decontaminated and more than 500 km of highway complexly deactivated. 700,000 transportation units were decontaminated, whilst 60 km of railway track and adjacent stations were also processed. However, decontamination activities have often been chaotic and inefficient, especially in the resettlement zone. In the Kyiv and Zhytomyr regions, for instance, temporary radioactive waste dumps were established which did not meet modern specifications and health norms.

A social dimension of nature conservation in the wake of Chornobyl was constituted by the emergence of public and political environmental campaigns. This coincided historically with the drive for *perestroika* and *glasnost* and saw environmental and related information distributed among the population. The environmental movement which previously existed on the territory of the former USSR, began to assume aspects of a political socio-ecological channel of protest against official environmental policies. The first years after Chornobyl (1987–88) should be considered the most active ones regarding a general sense of environmental commitment. In this period mass environmental movements and groups enjoyed unparalleled popular support.

The disintegration of the USSR, the establishment of a separate Ukrainian state, the negative socio-economic situation, declines in the living standards of the population - these all became factors which dramatically changed the distribution of forces in the political sphere. The leaders of most environmental movements in Ukraine staged a partial and then a full retreat from the initial political declarations of environmental concern. The leadership of environmental movements was to an extent co-opted by the state to work in the bureaucratic apparatus, and to an extent merged with other political parties. Environmental priorities were replaced by economic ones after much lobbying by government economists. Proof of this became evident when Ukrainian Parliament cancelled a moratorium on the construction of atomic power stations which had been adopted after the Chornobyl catastrophe.

Analysing the situation, it should be noted that in contrast to the western democracies, the process of establishing a multi-party system and environmental parties in Ukraine has only been underway for a few years. The burgeoning of environmental movements was only possible because of their appeal to a broad mass of people with an interest in the environment, as compared to the appeal of other parties to specific ethnic or linguistic groupings. The popularity of the "green" parties accordingly still remains high today. Among parties held in high public esteem, the "Green Party of Ukraine" (Partija Zelenych Ukrainy) is the front-runner (Kremen' 1994, Nyebozhenko 1993). It is interesting to note that interest in this party is determined not so much by its activities as by the ideas which it expounds, which are highly rated by society after Chornobyl.

There are numerous non-governmental environmental organisations in Ukraine. Ideally, they should orient their activities more towards providing an independent voice of opposition to current economic policies, carry out monitoring and independent analysis, and report violations to international organisations. In practice, such groups now mainly concentrate on educational initiatives, carry out environmental studies, and utilise informal contacts in relations with government representatives.

The environmental movement is most highly developed in the region of the capital city of Ukraine, where there are more than 119 organisations operating. 90 % of those are situated in Kyiv. By contrast, only one group is registered in the Khmelnytsk region, two each in the Poltava and Kherson regions, and three each in the Rivne and Volyn regions. There is a tendency towards concentration of socio-environmental movements in big cities (predominantly in rayon and region centres), which correlates in general with the higher level of civic action in cities. Big cities are also synonymous with larger numbers of the professional intelligentsia, who are often at the forefront of civic action.

At national and regional levels, there are citizens action groups supporting the rights of people affected by Chornobyl such as "Spilka Chornobyl" (Chornobyl Association), "Dity Chornobylya" (Children of Chornobyl), the "All-Ukrainian Organisation of Chornobyl Invalids" and others. Where in Ukraine as a whole the proportion of organisations devoted to Chornobyl is 3.75 % (NGO Handbook for Ukraine 1996), in the Zhytomyr region this figure stands at 12.5 %, in the Kyiv region 7.8 %, and in the Chernyhiv region 4.5 %, i.e. it exceeds the average Ukrainian level. This suggests a correlation between public activity on Chornobyl issues and the degree of actual radioactive contamination of the areas affected.

Conclusion

In the wake of the Chornobyl catastrophe, a region has emerged within Ukraine which is characterised by an absence of conditions conducive to sustainable development – an area affected by extreme radioactive contamination. Conversely, other, cleaner areas have had to cope with a considerable number of resettlers from the contaminated zones with all the social and economic problems this implies, and this has affected their own potential for sustainability. Given this state of affairs, there is an urgent need for regional policy intervention based on scientific recommendations and public opinion and possessing a complex ecological and social character. Implementation of such policy will encourage transition of the contaminated areas into a more sustainable phase of regional development. The social consequences of radioactive contamination also need to be considered by policy-makers when elaborating a plan for sustainable development in Ukraine.

The issue of foreign financial assistance remains of great importance. It should be noted, however, that the channelling of funding directly to central government as at present is not being as effective as foreseen. Future planning of grant-awards and financial assistance would do better to target regional and small and medium-scale development and research projects. Special attention should be paid to the creation of information corridors, at governmental and non-governmental as well as regional, national and international levels.

The study of the territorial specifics of socio-economic processes after the Chornobyl catastrophe shows that the degree of intensity of the environmental factor and its influence upon society depends on the degree of contamination by radionuclides. The impact of other factors remains an important subject for further research.

Essentially, the present research needs to be continued at a more local level (within the most contaminated regions). This is important from the point of view of singling out Chornobyl among other factors influencing the socio-economic situation as well as from the point of view of re-cultivation of the contaminated areas.

References

Counterpart Foundation (1996) NGO Handbook for Ukraine, Kyiv

Kaletnik NN, Landin VP, Pasternak PS (1991) Radioactive contamination of the forests of Polissya in Ukrainian SSR and organisation of forestry there. Radio-ecological and legal aspects of land-use after the Chornobyl accident. SOPS UkrSSR, Kyiv, pp 324–328

Kremen' V, Bazovkin E, Mischenko M et al. (1994) Socio-political situation in Ukraine: present state and development trends. Polytolohichni Chytannya, No. 1, pp 22–73

Marinich AM, Gorlenko IA, Rudenko LG et al. (1990) Constructive-geographical foundations of rational nature use in Ukrainian SSR. Theoretical and methodological studies. Naukova Dumka, Kyiv

Ministry of Chornobyl, Kyiv (ed) (1996) National Report of Ukraine. Ten years after the Chornobyl accident

Mischenko M. (1993) Social perception of young resettlers from the Chornobyl zone. Oykumena, No. 2, 1993, pp 4–6

Nyebozhenko V (1993) Ukraine: Political Crisis Deepens. Polytolohichna Dymka, No. 1, pp 94–96

Soboleva N, Donchenko N, Zlobina O et al. (1996) Attitude of the population affected towards the future. Sozialni Naslidky Chornobylskoi katastrofy. Folio, Kharkiv, pp 104–122

Statistical Yearbook of Ukraine (1995) Technika, Kyiv

The Ecological Impact of the Chernobyl Catastrophe on Sustainable Development in Belarus

Valentin Yatsukhno and Ludmila Kozlovskaya
Belarussian State University
Minsk, Belarus

Introduction

One of the principal causes hindering a sustainable, ecologically safe process of social and economic development in the Republic of Belarus relates to the negative consequences of the Chernobyl disaster. These consequences affect all the main spheres of social activity both in the contaminated regions and in the republic as a whole, and are a grave problem for Belarus. The nature and extent of the damage inflicted by the accident at the Chernobyl nuclear power station have become destabilising factors for the socio-economic advance of the republic, characterised as they are by a drastic deterioration of the ecological situation and a rise in morbidity levels. The problems brought about by the Chernobyl disaster involve changes in territorial structures relating to the use of natural resources, forced migration from contaminated regions, the depressed morale and psychological state of those affected, and the utter transformation of their whole way of life.

Extent and Ecological Consequences of Radioactive Contamination

The catastrophe at Chernobyl nuclear power station was, on account of its scale, complexity, and long-time consequences, one of the worst catastrophes ever in the history of atomic energy utilisation. According to estimates, about 50 million curies of 22 different radioactive isotopes was emitted into the environment. The substances released had their origins both in the immediate nuclear reactions (e.g. I, Cs, Cr, Ru) and in the reactor's fuel elements (U, Pu). For ten days after the disaster at Chernobyl, radioactive contamination was registered throughout Europe (Table 1).

The largest area of contamination was in Belarus. Fall-out from the accident affected roughly 25 % of the country, or some 4,000 communities and more than two million people. The territory of Belarus suffered not only the most extensive, but also the most intense contamination.

Approximately 20 % of the republic's agricultural land and about 25 % of its woodlands are located in the area affected. In the zone of radioactive contamination, radiologically unsafe lands deemed unsuitable for the production of food under both national regulations and international ecological agreements have been identified and withdrawn from economic use. To year end 1996, the total area of such lands amounted to 264,000 hectares. The Polesky radiation ecology reserve with a total surface area of about 220,000 hectares has been installed on territory with levels of contamination above 40 Ci/sq.km or above 1,480 kBq/sq.m.

Table 1. Caesium - 137 (total) contaminated areas in European countries in 000 sq. km
Source: Izrael (1996).

Countries	Area (in 000 sq.km) contaminated above specified levels (kBq/sq.m)						% of contamination deposited in Europe (%)
	10–20	20–37	37–185	185–555	555-1,480	>1,480	
Belarus	60	30	29.9	10.2	4.2	2.2	33.5
Russia	300	100	48.8	5.7	2.1	0.3	23.9
Ukraine	150	65	37.2	3.2	0.9	0.6	20
Sweden	37.4	42.6	12.0	–	–	–	4.4
Finland	48.8	37.4	11.5	–	–	–	4.3
Bulgaria	27.5	40.4	4.8	–	–	–	2.8
Austria	27.6	24.7	8.6	–	–	–	2.7
Norway	51.8	13.0	5.2	–	–	–	2.3
Romania	14.2	43.0	–	–	–	–	2.0
Germany	28.2	12.0	–	–	–	–	1.1
Greece	16.6	6.4	1.2	–	–	–	0.8
Slovenia	8.6	8.0	0.3	–	–	–	0.5
Italy	10.9	5.6	0.3	–	–	–	0.5
Moldova	20	0.1	0.06	–	–	–	0.45
Switzerland	5.9	1.9	1.3	–	–	–	0.35
Poland	8.6	1.0	–	–	–	–	0.23
Estonia	4.3	–	–	–	–	–	0.08
Czech. Rep.	3.4	0.36	–	–	–	–	0.09
Slovak Rep.	2.1	–	–	–	–	–	0.05
Lithuania	1.2	–	–	–	–	–	0.02

For decades to come, the main source of radioactive contamination will be Cs and Sr (with half-lives of 30 and 28 years respectively) and Pu (half-life of 86 years). It has been established that Cs accounts for about 90 % of the total radioactivity emitted. Radionuclides present in landscape systems contain two physical-chemical forms. The bulk of isotopes occurring in the southern region (over 80 %) are referred to as 'hot particles'. As a rule these isotopes have an average density of 10 particles/m^3 and are concentrated in the immediate vicinity of the Chernobyl atomic power station up to a range of about 30 km. In other regions the level of contamination is lower, though the contaminated area is large due to radioactive fall-out through rainfall. These nuclides are present in molecular form.

Stationary observation of nuclide migration bears witness to the fact that as a whole the process has accelerated in the environment. This is a result of active assimilation of nuclides by plants, wind erosion, leakage of surface waters into deeper lying soil layers, isotopic redistribution through water currents and localisation in bottom sediments and coastal deposits. However, the greatest worry concerns the fact that practically all radioactive nuclides have been incorporated into the biological cycle, i.e. soil–water–plants–living organisms.

This in turn adds to the probability of radioactive penetration of human organisms and concomitant increases in the morbidity rates of those exposed to radiation.

For those living in the contaminated zones, a particular hazard is posed by what is known as "resuspended" material. Resuspension is a phenomenon linked to secondary transfer, as a rule horizontal transfer of radioactive nuclides, which we have dubbed "the ink spot effect". Resuspension is caused by natural phenomena like wind and water erosion. Human activity is also a major contributor to the resuspension of radioactive material, for example, through agricultural practices such as ploughing and harrowing, through various modes of conveyance, especially on unsurfaced roads, and through scrub burning in woodland areas. The main consequences of resuspension are twofold: radioactive material can be transferred to uncontaminated areas and it can be inhaled by humans, thus increasing their potential exposure to radiation.

What are the main factors conducive to radioactive nuclides entering the environment? Three principal aspects attract our attention. First, the retention of Cs^{137} for a long period of time in the upper 0–5 cm of soil of and of Sr^{90} in deeper layers. Second, the fact that, the vertical migration of practically all radionuclides increases as one moves further away from the station. Third, the migration capability of americium^{241}, a by-product of Pu^{241}, is higher than that of plutonium. The half-life in adjacent and remote zones for Cs^{137} is 24–27 years. In the case of Sr^{90}, the half-life in 0-5 cm of soil for adjacent and remote zones is lower, extending from 7–12 years. According to one forecast, up to the year 2006 the Sr^{90}, Cs^{137} and $Pu^{239,240}$ content in the 0–5 cm soil layer in the adjacent zone will make up 30–40 %, 60–70 % and 90–95 % of their respective potential. For the remote zone, the equivalent figures will be Sr^{90} 15–25 %, Cs^{137} 35–45 %, and $Pu^{239,240}$ 10–20 % (Konoplya and Rolevich 1996). In the aftermath of the Chernobyl power station accident, 1,73 million ha or 25 % of the republic's woodlands were within the radioactive contamination zone (Table 2). From the beginning of 1988, an increase of root inflow of Cs^{137} and Sr^{90} isotopes to the above-ground phytomass against the background of continuous self-purification of surfaces was registered. The forecast findings show that woodland contamination is set to increase and that the main channel for isotopic transfer to the wood layer will be root penetration. In the next 10–15 years the above-ground phytomass, 30-year-old pine plantations in particular, will accumulate up to 10–15 % of the total amount of Cs^{137} in large forests. The Cs^{137} content in woodland edibles such as mushrooms and berries (whortleberries, cranberries, wild strawberries) exceeds the permissible norms (dried mushrooms 3,700 Bq/kg, berries 185 Bq/kg) even in areas with levels of soil contamination as 'low' as 37–100 kBq/sq.m. Radioactive contamination of woodland foodstuffs that limits their usage can be expected for the next 30–40 years in areas with a contamination density of 150 kBq/sq.m or above (Parfenov and Yakushev 1995).

Radionuclidic accumulation in animals is equal to the radioactive contamination in a given area. The highest levels of radionuclide content in various types of fauna were observed during the first year after the accident. From 1988 the process of nuclidic concentration slowed down considerably, and in some groups of animals (fish, amphibians, small mammals) inhabiting the most contaminated territories the radioactive content has since fallen back to 1986 levels.

Table 2. Radioactive contamination of woodland and other plant communities
Source: Konoplya and Rolevich (1996).

Cs-137 contamination zones, kBq/sq.m	Coniferous species, 000 ha		Hard deciduous 000 ha		Soft deciduous 000 ha			
	Total	Pine	Total	Oak	Total	Birch	Aspen	Alder
37–185	762.8	691.8	73.2	67.0	280.7	170.1	18.4	91.5
185–555	188.8	170.2	19.1	17.6	71.7	43.7	5.4	22.3
555–1,480	90.1	80.6	9.0	8.3	34.0	20.8	2.8	10.2
> 1,480	24.5	21.1	2.1	2.1	9.2	5.7	0.8	2.5
Total	1,066.2	963.7	103.4	95.0	395.6	240.3	27.4	126.5

Termination of industrial activity in the contaminated areas had an impact on the structure of species and volumes of birds and game. Within the 30-km zone, considerable and steady growth in the number of populations of game species has taken place (Figure 1). Due to the abundance of food and the cessation of hunting, the wolf population has increased 4 to 5-fold. And in settled areas one can encounter species of woodland and open-area wildlife. As a whole, the number of rare animals in faunal complexes has increased.

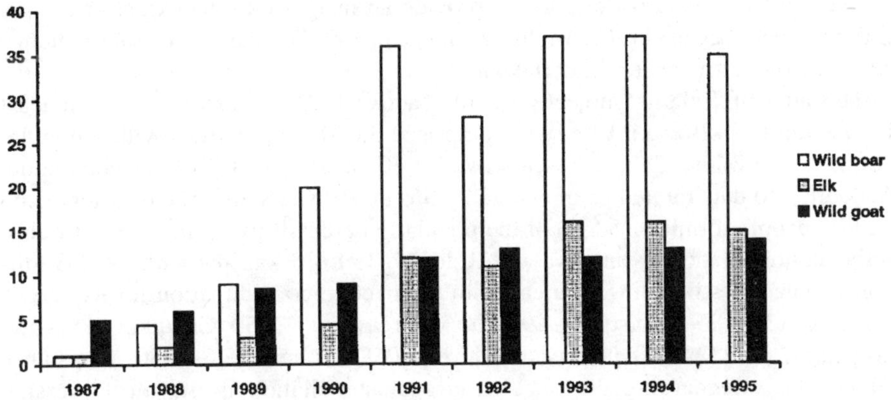

Figure 1. Population dynamics of selected animal species in the zone around Chernobyl (animals per 1,000 ha)
Souce: Pikulik and Nikiforov (1995).

The situation regarding parasites in the zone contaminated by radionuclides is worse than in control biocoenoses. Fauna and species of leeches and parasites living on wild birds and small mammals, or in their nests, in the radioactively contaminated areas are more abundant than in contiguous areas. It is to be expected that, in due course, the number of species of epidemic and epizootic significance in the contaminated areas will rise further (Pikulik and Nikiforov 1995).

The radiological situation in Belarus is characterised by complex, heterogeneous contamination by radioactive substances with varying half-lives, the presence of radioactive isotopes in practically all strata of ecosystems, and their intrusion into the geochemical and trophic cycles of migration.

Social and Medical Consequences

The after-effects of the Chernobyl accident on the sustainability of the process of social and economic development in Belarus are detrimental to both the present and future of the republic. This is due to the extraordinary extent of the damage inflicted.

Most severe are the social consequences, namely large-scale forced migration of the population (internal ecological migration) from the contaminated zone, the need to establish safe conditions for residents who remained in the contaminated areas, growing morbidity rates on account of radiation doses received, the social and psychological problems of adaptation to unaccustomed living conditions in the regions of radioactive contamination and new domiciles, growing rates of unemployment in the contaminated regions as a result of economic degradation of the latter and an aggravated crime situation.

The following data is designed to provide an insight into the extent of the social consequences of the Chernobyl accident in Belarus, their territorial structuring, and counter-measures undertaken.

The area of Belarus subjected to radioactive Cs^{137} contamination of at least 1 Ci/sq.km (37 kBq/sq.m) in 1986 comprised 3,600 communities with a population of 2.2 million. This represents over 20 % of the country's total population. According to data quoted as of 1 January 96 (i.e. 10 years after the disaster), 1.63 million people forming 15.8 % of the population are still living in zones of radioactive contamination (Konoplya and Rolevich 1996). More particularly, 1.3 million people are still living in areas with radioactive contamination levels of 1–5 Ci/sq.km (37–185 kBq/sq.m), 298,600 in areas with 5–15 Ci/sq.km (185–555 kBq/sq.m), and 24,400 in areas with 15–40 Ci/sq.km (555–1,480 kBq/sq.m). Given that safeguarding the population against radiation is the most pressing among measures aimed at overcoming the after-effects of the disaster, the Government of Belarus has been pursuing a programme of moving people from the most contaminated regions since 1986. Furthermore, special measures have been taken to provide safe living conditions for those who remained in the contaminated areas and social support for those who suffered from the accident (Table 3).

Table 3. Zoning of the territory of the Republic of Belarus by level of radioactive contamination and dose to which population was exposed

Name of the zone	Equivalent dose mSv per year	Contamination density, kBq/sq.m		
		Cs-137	Sr-90	Pu-238, -240
Residence zone with periodic radiation control	< 1	37–185 (1–5)*	5.55–18.5	0.37–0.74
Zone with evacuation rights	< 5 > 1	185–555 (5–15)*	18.5–74	0.74–1.85
Zone of later evacuation	> 5	555–1,480 (15-40)*	74–111	1.85–3.7
Zone of primary evacuation		(> 40)	> 111	> 3.7
Total exclusion zone	the territory around the Chernobyl NPP from which the population was evacuated in 1986			

By the end of 1986 more than 16,000 people from 108 communities had been evacuated from the 30-km zone around the Chernobyl nuclear power station, and by 1995, full evacuation of the population from areas with a contamination density above 40 Ci/sq.km and partial evacuation from areas with a contamination density between 15–40 Ci/sq.km had been practically completed. Relocation of a total of 131,000 people away from the contaminated regions had been organised by year end 1996. From 1989 to 1995, houses with a total floor space of 3,318 sq.m and more than 48 thousand flats were constructed for migrants by resorting to funds earmarked for dealing with the consequences of the accident. Expenditure on forced relocation during the period 1986–1995 is estimated to have totalled 4.36 billion USD.

Apart from organised relocation, population outflow from the contaminated regions was also to an extent initiated individually. It is possible to gain an idea of the magnitude of forced ecological migrations using data on population changes in different districts of Belarus over the past 10 years. Between 1985 and 1995, though the urban population in the republic as a whole increased by 14.3 %, in Gomel district it was only by 8.4 % and in Mogilev district by 6.8 %. During the same period the rural population of Belarus as a whole decreased by 13.6 %; the figure for Gomel district, however, was 23.1 % and for Mogilev district 14.7 %. In the 10 years referred to, total population increased by 3.7 %; in the same period, the population of Gomel district actually decreased by 4.2 % and that of Mogilev district by 1 %. Population decrease in the most contaminated areas of the Gomel and Mogilev districts averaged 46 % in the period from 1985–1995.

As has been demonstrated by sociological studies, the wish for a secure and ecologically sound life and for good health for one's children rates first among the reasons cited for induced migration. It can thus be inferred that the large-scale migratory processes of the post-Chernobyl period are forced in nature and have had negative material and psycho-social consequences for those moving.

Unfortunately, none of the measures undertaken (to wit urgent evacuation from contaminated regions and processes of compulsory and voluntary migration from radiologically hazardous zones that were dragged out for many years) were sufficient to ward off the appallingly deleterious impact of the Chernobyl disaster on the health of the Belorussian people. Morbidity and death rates have increased in Belarus due to the sizeable radiation dose received by practically everybody in the republic in the first days after the accident, the enormous dose suffered by 20 % of the population ("iodine shock"), and as a result of people being exposed to continued irradiation in contaminated areas (Figures 2, 3).

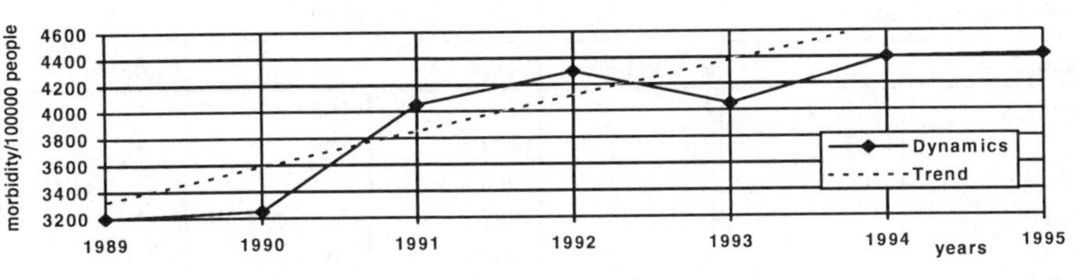

Figure 2. Morbidity dynamics and trends amongst those affected by the process of tackling the consequences of the Chernobyl NPP accident (per 100,000 people)
Source: Konoplya and Rolevich (1996).

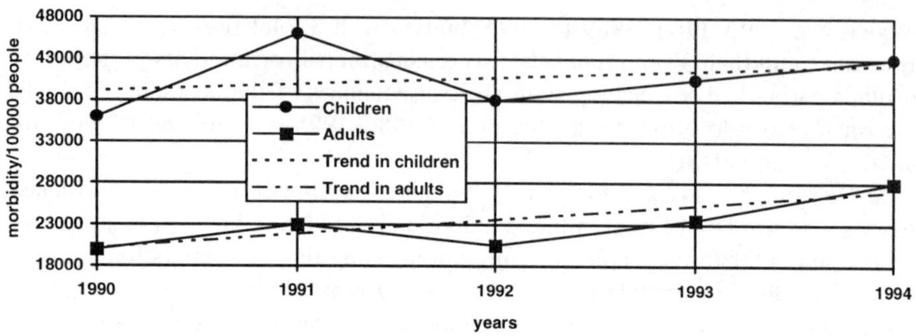

Figure 3. Morbidity dynamics amongst evacuees (per 100,000 people)
Source: Konoplya and Rolevich (1996).

Between 1985 and 1995 the birth rate in Belarus dropped from 16.5 % to 9.8 % (down by 40.6 %) whereas the death rate grew from 10.6 % to 13 % (up by 22.6 %). This tendency is particularly marked in Gomel and Mogilev districts. In the case of people moved from the area with a radioactive contamination level above 555 kBq/sq.m one observes a growth in the morbidity rate for all principal classes of diseases. Cases most often recorded relate to bronchopulmonary, circulatory, digestive, neural and endocrinal disturbances (Figure 4). Here, morbidity rates exceed the average for the republic by a factor of 2.1–9.8 (Konoplya and Rolevich 1996). At the same time, the number of cancer cases has also increased. Over the past 10 years the morbidity rate for malignant tumours in men has grown by 24.1 % and in women by 22.6 %. For those living in areas with a radioactive contamination level of above 555 kBq/sq.m, the rate has increased by 56.3 %, and at levels of between 185–555 kBq/sq.m the increase has been by 40.4 % (Table 4).

Table 4. Thyroid cancer incidence in Belarus
Source: Konoplya (1996).

Pre-accident period			Post accident period		
Years	Adults	Children	Years	Adults	Children
1977	121	2	1986	162	2
1978	97	2	1987	202	4
1979	101	0	1988	207	5
1980	127	0	1989	226	7
1981	132	1	1990	289	29
1982	131	1	1991	340	59
1983	136	0	1992	416	66
1984	139	0	1993	512	79
1985	148	1	1994	553	82
Total	1131	7	Total	2907	333

A link between an increase in malignant tumours of the thyroid gland and the after-effects of the Chernobyl accident has been established (Table 4).

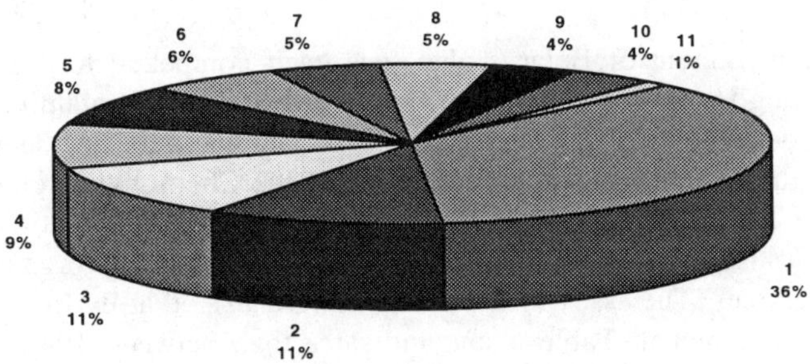

1 respiratory organs
2 blood circulation system
3 nervous system and sense-organs
4 digestive organs
5 bone and muscle system
6 endocrine system
7 infectious and parasitic diseases
8 urogenital system
9 skin and subcutaneous cellular tissue
10 psychosomatic
11 malignant neoplasms

Figure 4. Breakdown of the morbidity of those inhabiting areas contaminated with over 555 kB/sq.m of Cs^{137}

Serious disturbances of a psycho-emotional nature are also common amongst those affected by the Chernobyl disaster (e.g. evacuees, migrants, those who remained in zones of radioactive contamination and those involved in the clear-up). All these people were traumatised and are in need not only of medical observation, treatment and sanitation but also require social and psychological rehabilitation.

Economic Losses

Assessment of the financial damage caused by increased morbidity in Belarus during the post-accident period presents a problem, since many phenomena associated with genetic or other medical and biological after-effects of continued exposure to low radiation doses are unexplained and call for extensive investigations. However, approximate calculations, performed by Belorussian statisticians with the use of special procedures, have demonstrated that the total financial loss resulting from the deterioration of health caused by radiation doses received or expected and from psychological trauma experienced in zones of radioactive contamination over the period 1986–2015 can be set at 821.3 million USD. Taking expenditure on required sanitation measures as well as national income shortfalls on account of a high sickness rate into consideration, the aggregate loss arising from health impairments in the thirty years after Chernobyl is estimated at 1.9 billion USD.

In addition, the government of Belarus also sees itself compelled to expend large amounts providing for ecologically viable living conditions in contaminated regions. Outlay includes compensation payments for the victims of the accident as foreseen by the 'On the Social Protection of Victims of the Chernobyl Accident' Act, improvement of the social service system, radiological decontamination in contaminated areas, utilisation and elimination of radioactive waste and organisation of monitoring services. The assessment of expenditure arrived at by Belorussian statisticians is contained in Table 5 and indicates that, between 1986 and 2015, actual financial losses combined with additional expenditure aimed at minimising the social consequences of the Chernobyl disaster will amount to as much as 146 billion USD. Clearly, such expenditure cannot be borne by the Belarus economy, which is in a critical enough state as it is.

Heavy financial losses were inflicted by the accident on the whole sphere of material production (industry, agribusiness, civil engineering, forestry, transport and communications, utilities) as well as on natural resources (Table 5). Worst hit were branches of the economy based on natural resources and landscape utilisation, viz. agriculture and forestry. In Belarus 1,8 million hectares of agricultural lands, or 20.8 % of the entire area given over to agriculture in the country, was subjected to radioactive contamination. 264,000 ha of agricultural lands were withdrawn from the agricultural cycle due to high levels of radioactive contamination. Moreover, expert assessments show that output of unpolluted agricultural products is impracticable over an area of 800,000 ha. The total loss sustained by Belorussian agriculture is assessed at about 70 billion USD and includes direct losses from withdrawal of land from the agricultural cycle, the value of products not obtained due to resultant under-production, losses in productive and intangible assets, additional expenditure relating to the specific conditions of agricultural production pertaining in the contaminated areas (respecialisation in plant growing and cattle breeding, adoption of specialised methods and procedures, etc.).

In zones of radioactive contamination great losses are recorded in industries that process agricultural products. Processing enterprises have been obliged to close and move to non-contaminated areas (direct losses) or else have been involved in additional outlay on decontamination of equipment, the search for new suppliers of raw materials, radiological control etc.

As a consequence of the Chernobyl disaster, about 2 million hectares of woodland, or 25.3 % of the republic's entire forestry resources, are now within the zone of radioactive contamination. This fact has given rise to direct losses of timber, wood pulp, mushrooms, berries and other important resources, loss of the protective and recreational functions of woodlands, and additional expenditure on the decontamination of forestry products. The total loss to forestry is estimated at 4.1 billion USD for the 30-year period after the accident (Table 5).

Table 5. Estimates of the economic damage from the Chernobyl catastrophe for the Republic of Belarus, billion USD

Type of damage	Total damage per period				
	1986–1990	1991–1995	1996–2000	2001–2015	1986–2015
Health impairments	0.69	0.29	0.24	0.65	1.87
Industry	0.06	0.13	0.11	0.33	0.63
Social sphere	2.77	5.00	2.04	4.43	14.23
Construction	0.15	1.25	0.32	0.96	2.68
Transport & communications	0.93	1.20	0.36	0.90	3.39
Housing & communal services	0.07	0.45	0.92	2.02	3.46
Agribusiness	18.31	20.00	15.60	18.10	72.00
Timber industry undertakings	0.58	0.68	0.70	2.15	4.11
Resettlement	2.80	1.56	.39	0.33	5.08
Implementation of the Social Compensation Act for Victims of the Chernobyl Catastrophe	0.56	14.92	17.50	53.34	86.32
Pollution of mineral & water resources	2.00	0.12	0.15	0.10	2.67
Decontamination of polluted areas	0.04	4.19	22.48	10.12	36.83
Setting up of monitoring service	0.05	0.21	0.19	1.27	1.72
Total	29.00	50.00	61.00	95.00	235.00

340 industrial enterprises, which together generated 17 % of overall output in 1990, are located in zones of radioactive contamination. Analysis shows that the crisis phenomena that dogged of Belarus economic policy between 1990–1995 are most profound in the industries within the Chernobyl zone. There the setback in production is greatest on account of the closure of industrial facilities and the phasing-out of products.

Losses incurred by Belarus industry as a result of the accident at the Chernobyl nuclear power station are assessed at 630 million USD. The greater part of this (94 % of total value) is made up of losses caused by production cut-backs or curtailment.

The loss incurred by the building and construction, communications, and housing and communal sectors, relates to direct losses taking the form of equity capital relinquished in the exclusion and relocation zones, lost income potential due to the curtailment and cessation of production as well as additional expenditure on decontamination procedures and relocation and the setting-up of new economic units.

The aggregate loss incurred by the Belorussian economy in the wake of the accident and assessed at 235 billion USD for the period from 1986–2015 (or 32 annual budgets of the republic from 1985) is a prime cause of the fragile, unsustainable social and economic evolution as a sovereign state. Unfortunately, this tendency is likely to persist for quite some time to come. Expenditure allocated by the Government to tackling the consequences of the accident at the Chernobyl nuclear power station is placing a heavy burden on the national budget despite the fact that such expenditure is currently in a state of enforced curtailment due to the crisis phenomena associated with transition.

By way of example, budgetary funds earmarked for the Chernobyl programme in 1991 represented 16.8 % of the overall budget, whereas by 1995 these funds had been reduced to 7.3 % and by 1997 to just 6.3 %.

Considering the global ramifications of the Chernobyl disaster and the fact that it occurred beyond the borders, and through no fault of, Belarus and that Belarus nevertheless took the brunt of the impact of the accident and is, moreover, a small state with a stagnant economy, it can be stated that the government and the people of Belarus have a right to expect active support from the world community in overcoming the negative after-effects of what was, after all, the greatest nuclear disaster of the century.

References

Izrael YuA et al. (1996) The Atlas of Caesium-137 contamination of Europe after the Chernobyl accident. In: The radiological consequences of the Chernobyl accident. Proceedings of the first international conference. Minsk, Belarus, 18 to 22 March 1996. ECSC–ES–EAEC, Brussels-Luxembourg, pp 1–10

Konoplya EF (ed.) (1996) Ecological, medical, social and economical Chernobyl catastrophe consequences in the Republic of Belarus. Minsk, 155 pp

Konoplya EF, Rolevich IV (eds) (1996) The Chernobyl catastrophe Republic of Belarus The national report. Minsk, 96 pp

Parfenov VI, Yakushev BI et al. (1995) Radioactive contamination of vegetation of Belarus connecting with Chernobyl accident. Nauka i technica, Minsk, 582 pp (in Russian)

Pikulik MM, Nikiforov ME et al. (1995) The animals within the zone of Chernobyl accident. Nauka i technica, Minsk, 263 pp (in Russian)

Sustainable Development of the Regions in Ukraine

State of the Economy and Development Perspectives

Leonid Rudenko and Inga Horlenko
Institute of Geography, National Academy of Sciences
Kyiv (Kiev), Ukraine

Economic Development Trends in Ukraine

Introduction

At the end of the second millennium Europe is experiencing the strengthening of both integrational and disintegrational processes, and this is having a significant impact on the development of its regions. The collapse of the Soviet Union and the Council for Mutual Economic Aid (CMEA) resulted in new independent states appearing on the world map, and these and other states of the former CMEA have radically altered the course of their social and political development. The economic system in the countries of eastern Europe is changing along with their internal markets and foreign economic relations. This in turn is exerting an influence on the system of international relations. The economic and political reforms in the countries of eastern Europe have been accompanied by a significant fall-off in production, deteriorating living standards, growing unemployment and increasing social unrest.

In Ukraine it was expected that the transformation process would pass through the following four stages. The first stage of the transformation model was to be a preparatory phase of economic development planning (e.g. outlining schemes for the liberalisation of national and foreign commerce, liberalisation of prices, elaboration of privatisation programmes, and expansion of the service sector with particular emphasis on financial services). The second stage was to be characterised by measures of political and legal transformation including public awareness campaigns. The practical transformation was not to start until the third phase. This would be the most difficult stage because of the emergence of such phenomena as inflation, unemployment, the altered property status of enterprises, enterprises closing and becoming bankrupt, the restructuring of the economy, the adoption of new internal and external functions by enterprises and regions, etc. The general transformation process was to be consolidated in the final stage, with the economy in the regions being stimulated and standards of living improving. An essential feature of development at this stage would be the right distribution of investment throughout the Ukrainian regions in the interests of decentralisation of production and more responsibility for all producers. Unfortunately, Ukraine has not negotiated these four stages one by one but rather all at once and in an often wholly uncoordinated manner.

As one of the former Republics of the Soviet Union, Ukraine has undergone momentous political and economic change that has inflicted great hardship on the Ukrainian population. The main factors adversely affecting economic development in Ukraine since the declaration of independence in 1991 are the following:

- sudden transition from a centrally planned economy to the principles of market economy management without the establishment of the respective legislative base and with a lack of specialists able to put these principles to effect at the different levels of authority;
- hyperinflation against a background of contracting production, escalating unemployment, and 'short-time working' has hit the population hard.

Development of Major Economic Indicators

After five years of transition the Ukrainian economy has yet to be consolidated. Overall gross domestic product has declined by a factor of 2.5, industrial production has halved, agricultural production is a third down and capital expenditure has shrunk by a factor of four (see Table 1). Production in the leading branches of the economy has been declining from year to year. The industrial sector with the biggest decline is light industry with 47.4 % in 1994 and 34.8 % in 1995 (Table 2). The same dramatic trend was also witnessed in 1996. For the first 9 months of the year production in light industry declined by 24.6 % (and in the first quarter by 32.2 %).

Output in the textile industry declined by about 40 % in 1995.

Table 1. Development of major economic indicators 1991–1995 (% to previous year)

Indicator	1991	1992	1993	1994	1995
Gross Domestic Product	91.3	90.1	82.9	77.0	88.2
Gross National Product	92.5	89.8	92.1	74.8	87.3
Agriculture	86.8	91.7	101.5	83.5	96.0
Industrial production	95.2	93.6	92.0	72.7	88.5
Food industry	87.0	86.2	86.7	81.5	86.2

Source: Economics of Ukraine, 1996, vol 7, p 20

Table 2. Decrease of production in selected branches in Ukraine (% to previous year)

Branches of economy	1992	1993	1994	1995
Electricity	-7.1	-3.6	-12.5	-5.4
Petrol	-13.0	-23.1	-17.2	-10.5
Ferrous metallurgy	-10.1	-21.4	-28.2	-6.6
Non-ferrous metallurgy	-16.9	-10.5	-24.7	-2.4
Engineering industry	-5.8	4.3	-41.2	-24.8
Light industry	3.1	-15.2	-47.4	-34.8
Food industry	-16.8	-11.5	-19.7	-13.8
Building materials	-8.8	-18.9	-35.7	-29.1

Source: Ukraine in Numbers, Kyiv, 1996

Similar reductions have been registered in food production. In 1996 the production of meat declined by 35 %, while 40 % fewer sausages and 46 % fewer dairy products were produced. The situation in many other branches was similar.

As a first step towards a national currency, in 1991 the Ukrainian government introduced 'coupon-karbovanets' in place of the Soviet rouble. In 1996 a new Ukrainian currency, the 'grivna', was created. The coupons had very rapidly depreciated in value. Hyperinflation has been stoked by rising prices charged by the energy utilities, which in 1990 were about 10 % of world prices. After independence, the price of oil and gas in Ukraine moved towards parity with world market prices. The average monthly inflation rate peaked in 1992 (Table 3). Table 4 shows the monthly inflation index for consumer prices.

Table 3. Inflation in Ukraine

	1991	1992	1993	1994	1995
1. Inflation rate over previous year (%)	290	2.7	10.1	401	182
2. Average monthly inflation rate (%)		77.4	49.7	19.2	9.2
3. Exchange rates (karbovanets for USA $1 at year end)		749	25,000	104,000	179,400

Source: Ukraine in Numbers, 1995, Kyiv

Table 4. Monthly inflation (Index of consumer prices)

	1992	1993	1994	1995
January	252.0	73.2	19.2	21.2
February	22.8	28.8	12.6	18.1
March	10.4	22.1	5.7	11.4
April	13.7	23.6	5.9	5.8
May	22.8	27.6	5.2	4.6
June	18.0	71.7	3.9	4.8
July	27.5	37.6	2.1	5.2
August	15.6	21.7	2.6	4.6
September	149.0	80.3	7.3	14.2
October	25.4	66.1	22.6	9.1
November	22.6	45.3	72.4	6.2
December	38.2	90.8	28.4	4.6

Table 5. Food supply in Ukraine (1994, kg per person)

Foods	Physiological standard for consumption	Actual consumption	Deviation of actual consumption from physiological standard	
			kg	%
1 .Meat and meat products	83.0	43.5	-40.0	-48.2
2. Milk and dairy products	395.0	254.0	-141.0	-35.7
3. Eggs (units)	291.0	181.0	-110.0	-38.8
4. Fish and fishery products	20.1	2.9	-17.2	-85.6
5. Sugar	38.0	36.0	-2.0	-5.3
6. Sunflower oil	13.4	8.0	-5.4	-40.3
7. Potatoes	124.0	130.0	+6.0	+4.8
8. Vegetables and gourd crops	161.0	82.5	-79.0	-49.1
9. Fruits and berries	90.0	24.0	-66.0	-73.3
10. Cereal products	101.0	140.0	+39.0	+38.6

Source: Economics of Ukraine, 1996, vol 7, p 23

Logically enough given such inflation, bank savings were devalued, many enterprises collapsed, unemployment reared its head, part-time working was introduced (as often as not without any real work content) and general standards of living dropped (Table 5). About 80 % of the Ukrainian population cannot subsist on salaries alone but depends on other sources such as after-hours gardening at the traditional datchas.

Table 6 shows the development of the exchange rate in Ukraine for the period 1992–1995. In 1996 media reports on issues such as the inflation process and production became more optimistic. Unfortunately, this does not reflect any real improvement, since the fall in inflation is not the result of economic stabilisation but rather of delays in the payment of wages, pensions and grants and, naturally enough, through the taking out of new loans. The delay in paying salaries varies by industry from 4 to 6 months.

Table 6. Development of the exchange rate in Ukraine (Ukrainian karbovanets to 1 US-$)

Year	Month		Rate of National Bank		Rate for Interbank Currency Market		Unofficial Rate	
1992	January	December	0.56	637.7	110	749	130	1050
1993	January	December	931	12,610	1,002	25,000	1,600	37,000
1994	January	December	12,610	104,200	25,000	104,200	39,500	133,000
1995	January	December	117,200	179,400	117,200	179,400	145,000	185,000
1996	January		187,500		187,500		192,000	

Foreign investment has a crucial bearing on the further course of transition, but given the inability to finalise taxation laws, the strength of the bureaucracy, and the convoluted and shifting legal framework in place for private enterprise (e.g. difficulties in registering and licensing foreign investment), foreign investment in 1992 and 1993 was negligible. Levels did not increase materially in 1994 and 1995 either, totalling US$ 483.5 million and US$ 750.1 million respectively (Table 7).

That the Ukrainian balance of trade for 1995 was positive is only due to high exports of agricultural products (Table 8), which in return aggravated difficulties in the national food supply chain.

Table 7. Foreign investment in Ukraine

	1994		1995	
	million US$	%	million US$	%
Germany	101.3	20.9	129.6	17.3
USA	96.6	20.0	171.4	22.9
Great Britain	33.8	7.0	46.3	6.2
Cyprus	28.5	5.9	38.6	5.1
Switzerland	21.3	4.4	35.2	4.7
Canada	14.8	3.1	17.7	2.4
Russia	19.1	4.0	37.8	5.0
Poland	13.1	2.7	18.9	2.5
Hungary	13.7	2.8	19.5	2.6
Total	342.2		515.0	
Overall international investment	483.5		750.1	

Source: Ministry of Statistics (1995) Kyiv

Table 8. Ukrainian balance of trade (1995, by general sector, million US$)

		Balance of trade with countries from CIS	Balance of trade with other countries
Overall balance	-1,202.0	-2,308.3	1,106.1
Energy (incl. fuel)	-7,138.7	-6,511.6	-627.1
Agricultural goods	1,633.0	1,639.6	6.6
Light and local industries	-183.5	-98.8	-84.7
Engineering and metal-working industries	-742.7	341.5	-1,084.2

Source: Ministry of Statistics (1995) Kyiv

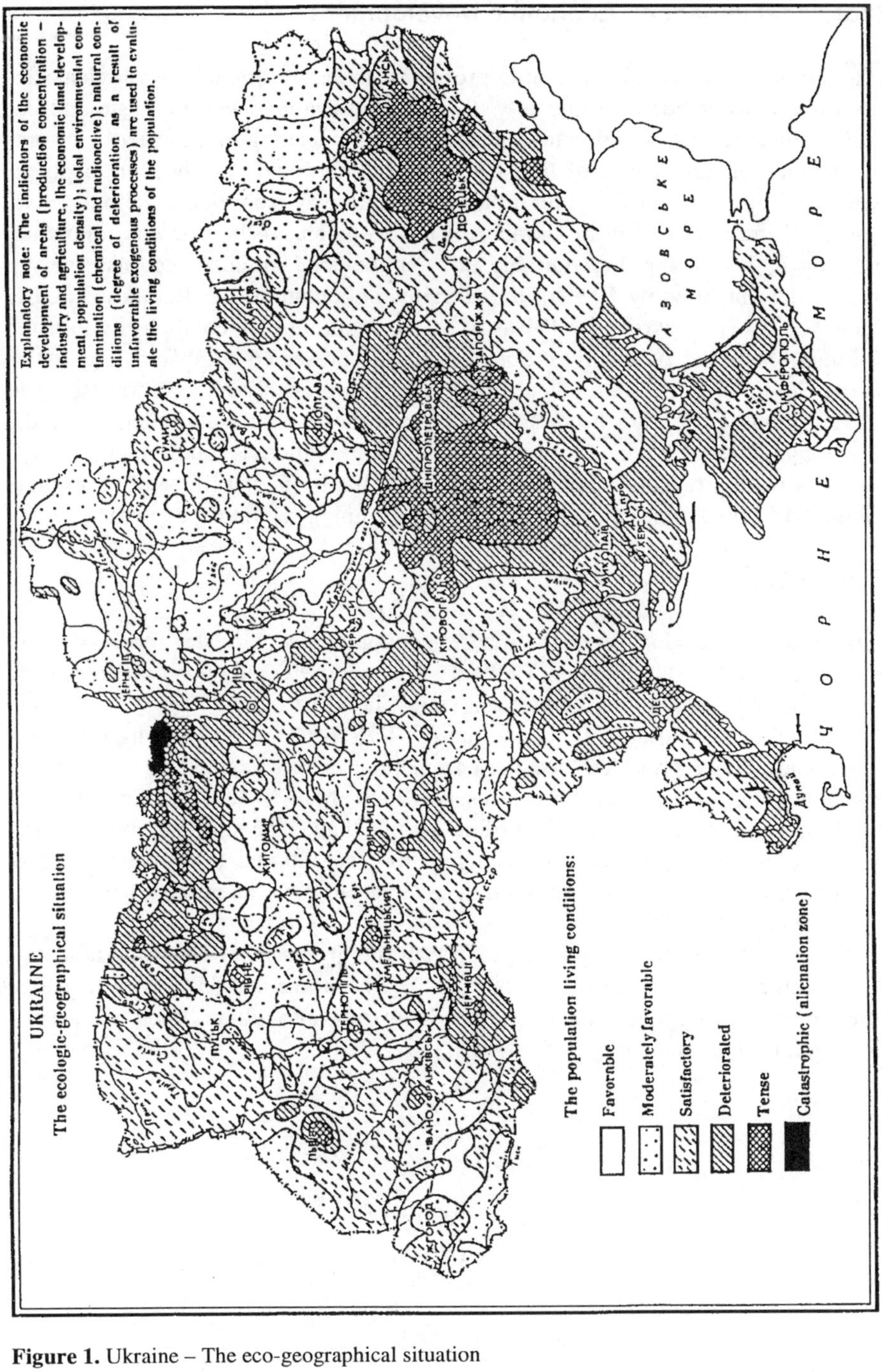

Figure 1. Ukraine – The eco-geographical situation

Ecological Impact on Economic Development

In addition to specifically economic problems, there are dramatic ecological factors to be considered. Unfortunately, there is a still a tendency to waste and degrade natural resources and landscapes, despite this leading to a further deterioration in the general quality of life. Chornobyl (Chernobyl) has become a potent symbol of environmental calamity, but it is a long way from being the only environmental problem region in Ukraine. It has to be stated that the environmental situation is developing unfavourably in a number of Ukrainian regions, over an area in fact inhabited by 70 % of the Ukrainian population. VA Baranovskyi, LG Rudenko, IO Horlenko and VP Rasov of the Institute of Geography within the Ukrainian National Academy of Sciences have represented comparative statistics on the Ukrainian eco-geographical situation in comprehensive map form (see Fig. 1). Enormous efforts are needed in Ukraine to cope with the challenges of the Chernobyl disaster. Besides environmental pollution, there is the problem that the country's natural resources are at the stage of exhaustion, and this will necessitate a large number of mines closing in the Donbas mining region for example.

Privatisation

Parallel to the introduction of the market economy, new (private) forms of property were introduced in Ukraine. Between 1992 and 1995, 28,000 companies were privatised or accorded alternative legal status, half of them in 1995 (Government of Ukraine 1996, p 6). At the beginning of 1996, 65 % of all catering establishments and service providers became commercial enterprises and 12 % of industrial companies became private entities, of which about 10 % were agricultural or building enterprises. About 15 million people have become shareholders and owners of private property. According to the Institutional Reorganisation Programme, the process of privatisation is assumed to have been completed by the end of 1996 (Table 9).

As a result of introducing new property forms it is assumed that the non-state sector will grow to comprise 75 % of industry as a whole and 95 % or even 100 % of enterprises in light and manufacturing industries, agriculture, trade and service provision (Government of Ukraine 1996, p 6).

Table 9. Privatisation in Ukraine (to the end of year)

Year	Number of private enterprises	Number of private farms	Number of citizens using their property certificates (000 persons)	Number of privately owned apartments (000)
1992	30	14,681	5	-
1993	3,585	27,739	728	902.8
1994	11,552	31,983	7,106	1,812.3
1995	19,800	34,149	21,228	2,236.3

Source: Economics of Ukraine, 1996, vol 4, 1996, p 11

Research and Development

Naturally enough, the economic decline has been accompanied by a vigorous reduction in funding for research and development. In 1990 the R & D budget was equivalent to 3.4 % of Ukrainian GDP, whereas in 1995 it stood at just 0.5 %. The reduction could well have major repercussions for Ukrainian development in general and specifically for the country's economic development, since research and development are vital preconditions for the generation of new products. Systematic delays in the payment of salary to National Academy staff and those at other research institutions have effectively caused key scientific capacity to drop by more than 40 % over five years. Currently, average outlay on scientists is US$ 700 per annum in Ukraine (compared with US$ 150,000 in USA and Japan). Clearly, given delays in payment, a smaller scientific workforce, and reduced budgets for the technical research infrastructure, Ukraine will dwindle in importance as a research centre and Ukrainian research will not be able to satisfy the nation's needs for new technologies and know-how. Ukraine will have to spend even more money purchasing new technologies and new equipment, while the brain-drain could further increase.

Conclusions from the First Part of the Foregoing

The protracted depression in the Ukrainian economy is marked by very difficult, debilitating and disintegrating processes at the various territorial and administrative levels, developing against the background of inflation. For more than half a century Ukraine was part of the Soviet Union and, together with the other countries of the CMEA, formed a common economic space that is now flailing. These processes have led to drastic cuts in production, an imbalance in regional development, and an avalanche of economic and, in turn, social problems. After a series of political and social volte-faces, the influence of national planning on the processes of transformation has gradually waned. There is now a need to implement an effective regional policy with an improved system of regional administration.

Perspectives for Sustainable Regional Development in Ukraine

Is it appropriate to discuss sustainable development under the present conditions, or, conversely, is it rather more important than ever that this be done? This question was discussed from a predominantly theoretical and scientific-methodology point of view during the Conference of the Network of Spatial Research Institutes in Central and Eastern Europe held at Vilnius in December 1995 (Rudenko, Horlenko 1996).

In Vilnius we were already considering more practical aspects, however, and presented proposals regarding what should be done next and also a strategy of sustainable development. On 15 October, 1995 the Supreme Soviet of Ukraine debated an Action Programme for the New Cabinet of Ministers, to be enacted in the light of the very difficult economic and social conditions and which was adopted in late 1996.

It was at the Sixth Conference of European Ministers responsible for regional/spatial planning, held at Torremolinos in Spain in May 1983, that the European Regional/Spatial Planning Charter was adopted (Council of Europe 1983). As stated in the concept adopted, regional/spatial planning gives geographical expression to the economic, social, cultural, and ecological policy of society. The well-being of all people and their interaction with the environment are central to regional/spatial planning. Regional/spatial planning has the declared objective of seeking at one and the same time to achieve balanced socio-economic development of the regions, improve the quality of life, responsibly manage natural resources and conserve the environment and make effective use of land.

Showing a certain sensitivity to the fate of the regions, the Ukrainian Prime Minister has laid emphasis on developments towards regional self-sufficiency as forming the main thrust of government policy. For Ukraine the most important function needing to be fleshed out and lent detailed definition is a regionalisation of socio-economic development to reflect territorial variations in natural and environmental conditions, natural resources, and economic and scientific potential. Regionalisation assumes substantial scientific differentiation between regions and the elaboration of a strategy of socio-economic development for each region that takes into account its distinctive characteristics and the interests of its population and will find support at national level.

Exactly what form regionalisation should take is still an open question. Various authors distinguish between 6 and 10 regions in Ukraine, interspersed with economic areas. A generally acceptable regional scheme is still lacking. However, it can be assumed that this will change, and in at least one or the other region regional planning is on the horizon.

The strategy of distant perspective in sustainable development both for regions in Ukraine and across Europe is rooted in the wish to balance the requirements of society with the need for a sustained development of our physical environment. The strategy of close perspective, by contrast, involves stabilising the economy in the regions and outlining practical ways of effecting transition processes while adhering to the principles of the renaturalisation and conservation of nature as well as upholding the balance between Society and Nature. Equilibrium presupposes the "sustainability" of all regions. To the general balance may be added such between:

- the specific natural resource potential of an area and structures, the internal and external functions of production;
- the level of technogenic load and the powers of resistance of natural landscapes;
- levels of economic development and living conditions/reproduction;
- the territorial structure of production and the ecological framework of an area;
- scales of technogenic impact and biota facilities to counteract damage;
- incidence of risk situations developing and measures for their neutralisation.

The process of transition towards sustainable development in the Ukrainian regions is slow and cumbersome. It is always going to be complicated and besides the economic dimension also has ethical and social implications. Much effort will be invested in explaining to the population the vulnerability of the biosphere, in urging people to care for resources, in conserving and sustaining the quality of the individual constituents and biological diversity of nature and in conducting measures to renaturalise nature.

Given the current economic situation, the transition period requires strategies on regional development to be coordinated with national goals; production structures to be aligned and in compliance with the overall capacity of an area to aid rational utilisation; development to be geared towards the most effective social and natural balance (preservation and improvement of the quality of life); inter-regional links and trans-frontier cooperation to be activated; active participation of the community in certain strategies of regional development. It is especially necessary to highlight what is needed to restore and develop Ukrainian territorial planning, to wit a mixture based on constitutional principles, 'good traditions' from our own country and some experience from western Europe. It appears to be necessary, for example, to draw on the experience of the Commission for Territorial Planning operating within the Committee on Regional Affairs in the European Parliament.

The processes of economic transformation taking place in Ukraine need to acquire a more active regional component on foot of the regions recognising the freedom thus brought about to tackle both intra- and interregional problems. The transformation of the Ukrainian space presupposes changes in the system of intra- and interregional proportionalities and relations, the emergence of new functions, a sense of balance between the centre and the periphery of a region, and between the region itself and the national capital, and shifts in the geopolitical standing of the regions. It is reasonable to pay more attention to problems at microlevels and hence to focus on Ukrainian subregions, which may be considered from a certain point of view as the "locomotives" of growth and development of the regions.

As regards the activation of transformation processes, it is important to strive for targeted spatial development based on the principles of renaturalisation of certain areas, conservation of the natural and cultural heritage, and reinforcement of popular awareness and morale. It is reasonable to elaborate a detailed transition model for sustainable development, implement processes of adaptation, train the public in implementation of this development model (nature conservation, careful use of resources, prevention of environmental pollution, etc.), and develop methods and an infrastructure for planning and implementing spatial development.

All efforts towards the development aspired after are to be based on democratic principles embracing all strata of society, and at the same time this development process has the task of achieving a balanced development of the regions so as to include Ukraine in the world-wide drive to eliminate social, ecological and other conflicts.

To move from vision to action, i.e. to put to effect the vision of sustainable development of the regions of Ukraine in the near future, a suitable strategy is necessary. The aims of this strategy should be as follows:

- to outline concepts for processes of economic transformation in the regions;
- to remedy the imbalance between natural and industrial-agricultural land and in this context reflect and, with respect to future development, prevent excessive land use in the economic development process;
- to implement phased reclamation of topsoil to prevent further losses of humus, decrease soil erosion and raise its productivity;
- to intensify work on the conservation and reintroduction of certain kinds of natural resources, the extension of biodiversity and expansion of the network of nature conservation areas in a national/regional ranking system;
- to deploy across the board new technologies designed to capture, confine, and eliminate noxious gases and solid and liquid industrial wastes;
- to explain to the public at length that there is no alternative to sustainable development for the regions.

References

Government of Ukraine (1996) Economy of Ukraine for January–September 1996, Governmental Courier, 11.2.1996, Kyiv

Council of Europe (1983) European Regional/Spatial Planning Charter, Torremolinos Charter. Strasbourg

Popovkin VA (1995) Towards a National Concept of Regional Economic Policy, Kyiv

Rudenko L (1984) Cartographic Substantiation for Spatial Planning, Kyiv

Rudenko L, Horlenko I (1996) Towards a Sustainable Development in Central and Eastern Europe. In: Ulrich Graute (ed) Vision and Strategies around the Baltic Sea 2010 and their Relevance to Central and Eastern Europe, Dresden: Institut of Ecological and Regional Development, IÖR-SCHRIFTEN, vol 17, pp 30–34

Yekhanurov YU (1996) State of Privatization in Ukraine: Problems and Ways of Solution. Kyiv: Economics of Ukraine, vol 6, pp 4–13

Regional Development Concept for the Brest–Minsk–Orsha Section of the Berlin–Moscow Trans-European Transport Corridor

Victor Ivlichev and Isaac Khazdan
Institute for Regional and Urban Planning of Belarus
Minsk, Belarus

Introduction

The Berlin–Moscow trans-European transport corridor comprises the territory located along motorway and railway trunk routes between the two cities and being within their active zone of influence. There are 3.4 million people or 33.8 % of the total population of Belarus living in the urban and rural settlements of the corridor region (within the boundaries of 16 administrative districts traversed by the E 30 motorway and the E 20 railway). Located here are Minsk, the country's capital and largest economic and cultural centre, Brest, one of the most dynamically evolving oblast centres, and the three large industrial cities of Baranovichi, Borisov and Orsha. The degree of population concentration and the economic potential within this zone have for decades and indeed centuries supported a trend towards stable growth, mainly as a result of conditions conducive to economic development that have ensured the ongoing evolution of transport communications from east to west and vice versa.

Work currently underway or concluded in the region and funded by loans from the European Bank for Reconstruction and Development (EBRD) includes phased redevelopment of the Brest–Moscow motorway (E 30), a fibre-optics telecommunications link from Orsha to Berlin, and construction of a new transcontinental gas pipeline from Russia to Poland and Germany. The region has several airports that can accommodate any type of modern aircraft. The railway from Brest to Moscow via Minsk is being upgraded for speeds of up to 160 km/h. At the European level, there are firm plans to construct a new high speed railway (HSR) from Berlin to Moscow via Minsk as an integral element of an European HSR network with envisaged train speeds of up to 250–350 km/h. Implementation of these plans is helping pump-prime and stimulate a revitalisation of the economy.

The commitment of the world community to modernising and expanding trans-European transport communications passing through Belarus brings with it the need to define a strategy for the future development of the region, as does the economic and demographic potential of settlements located within the corridor zone and the crucial role this potential will play in shaping the economic progress of the republic as a whole. One of the key tasks when modernising the infrastructure of the region is to square the interests of Europe, the republic of Belarus, and the inhabitants of the individual communities.

Recognising the importance and complexity of this problem, a start was made in 1995 on elaborating a joint project with the goal of defining a strategy and programme for spatial development of the corridor region as foreseen by a resolution of the Belorussian-German Council for economic cooperation and backed by the German Ministry for Regional Planning, Construction and Urban Development and the Belorussian Ministry for Architecture and Construction. The main executors of the project are: BelNIIPgradostroitelstva (Minsk) and PLANCO Consulting GmbH (Essen). Finding application in the project are the profound experience of Belorussian specialists in the sphere of physical and urban planning under the specific conditions of the centrally planned economy and the regional planning methodology in common use in Germany; the project is striving for maximum involvement of all interested parties (central and local administrations, managers of the state owned enterprises, the business community and the public at large) in the discussion of project goals and objectives to ensure the most efficacious solutions are found.

The project's information base comprises data from official statistics, research and development work as well as the findings of field surveys and questionnaires, including a special survey of transport and economic links, motorway services, and cargo traffic passing border crossing points.

Analysis in the first stage, in 1995, focused on problems of spatial development within the zone of influence of the trans-European corridor, with a preliminary forecast being made regarding the impact on the spatial organisation of the settlement network and natural resources of a rehabilitated international transport infrastructure.

In 1996 a general concept for spatial development of the region was devised that focused on harnessing the advantages of modernisation of the region and the envisaged expansion of the international transport infrastructure. Taking the example of Minsk, an attempt was made to flesh out this concept for one of the main development centres of the region; work also started on setting up a GIS for the region for the benefit of potential investors. In the final stage of the project, in 1997, it is planned to draw up detailed top-priority urban construction programmes for the main development centres of the region and to determine the conditions for implementation of these programmes.

The concept has been elaborated by the following authors:

<table>
<tr><td colspan="2">BelNIIPgradostroitelstva</td><td>PLANCO Consulting GmbH</td></tr>
<tr><td colspan="2">Isaac Khazdan
(head of Belorussian group of specialists)</td><td>Holger Platz
(team leader on the German side)</td></tr>
<tr><td>Tamara Butrimovich</td><td>Natalya Jurgenson</td><td>Uwe Samland</td></tr>
<tr><td>Leonid Desyatkov</td><td>Juri Chanturia</td><td></td></tr>
<tr><td>Igor Borisik</td><td>Semen Brandin</td><td></td></tr>
<tr><td>Georgi Baranetz</td><td></td><td></td></tr>
</table>

General project supervision on behalf of BelNIIPgradostroitelstva was undertaken by its director Mr. V. Ivlichev.

Prerequisites and General Goals of Spatial Development

The concept for the spatial development of the region occupied by the Berlin–Warsaw–Brest–Minsk–Moscow trans-European corridor (TEC) is mainly directed at producing a feasibility study on a variety of measures to ensure the maximum possible economic and social yield for the corridor region and the republic as a whole through modernisation and enhancement of the system of international transport and engineering communications and simultaneous neutralisation of possible damage to the environment.

The concept envisages step by step implementation of the "decentralised concentration" principle as being most conducive to the spatial and economic development of the republic, since almost half the population of Belarus lives in towns and cities.

The given economic situation for the near future will require the concentration of scarce state resources in the main development centres and industries, since these promise the quickest return on investment and play the role of a multiplier for the economy as a whole. The policy of prioritising promotion of the main development centres will enable these to preserve and enhance their economic potential so as later to provide stimuli necessary for the development of other areas (small towns, rural areas). Concentrating investment in the well-developed zones along the main transport axis of the republic will also contribute to solving a series of environmental and nature conservation problems.

In this respect the TEC region, which has historically been the main axis of economic development in the republic, has all the necessary conditions for implementing the principle of "decentralised concentration". The region and its main development centres have already become the catalysts for a great number of social and economic innovations.

A further concentration of resources within the TEC, where a highly qualified workforce is required, is quite logical from the point of view of both long-term development strategies for the country and the tactical issues of current transformations. However, *it is necessary* to rigorously *control the degree of concentration* so as to prevent negative consequences for the social and economic development of other regions that are allegedly being deprived of necessary resources.

In this context it is important that for all the cities of the region, irrespective of their own potential, the necessary prerequisites for infrastructure development are created, as if from outside, in the interest of other areas, or indeed states, i.e. that international transport and service links are established from east to west and vice versa. At the same time, it is necessary to continue the proven practice of encouraging small-scale production outside large centres which will provide input for the industrial complexes of these centres.

This will not only facilitate an increase in the economic potential of small and medium towns in the region but will also alleviate migratory pressures on large centres; it will allow deconcentration to be carried out more actively and the role of vast industrial districts in large cities to be modified, thus ensuring growth of jobs and changes in employment structures of their extensive zones of influence.

Regional potential and the main obstacles to putting it to effect are summarised in the following table.

Potential	Problems
There are good transport communications in the form of motorways and railways in the whole of the country in the direction of Central/Western Europe, Russia, Baltic states, Ukraine. Motorways and railways have sufficient throughput reserves, and the transport system as a whole has the potential for a considerable increase in traffic volume.	There are problems with respect to road surfaces, transport safety, permissible speed limits along motorways and railways, there is lack of transport and distribution services, lack of services for road users, weak interrelation between different types of transport.
The power distribution network is in an excellent state, this is also true of oil and gas pipelines.	There is a general deficit of locally generated electricity.
The telecommunications infrastructure within the corridor has been modernised.	The quality of domestic and international telephone calls remains low in some cities and districts
A number of the main cities, including the capital city Minsk, have potential for decentralised development (Brest, Baranovichi, Borisov, Orsha).	There is a tendency towards excessive concentration on Minsk; environmental conditions in the cities need improving.
In the corridor zone there is a high concentration of general and specialised educational establishments	Possibilities for mass education to new professions are limited, a high degree of specialist training within enterprises has led to overspecialisation of personnel
There are a lot of important industrial enterprises	Many of them operate at 50 % of productive capacity due to lost sales markets, high production costs and the slow rate of restructuring
There exist strong links with the huge Russian economy in the forms of raw materials processing and industrial semi-finished goods imported from there, re-export of processed products	Revival of Russian markets is yet to materialise

Many industrial enterprises are making enormous efforts to develop, design and manufacture new types of products and search for new sales markets	Information on what customers want and on approaches to marketing is lacking; foreign partners are displaying reticence, and the privatisation rate is slow; local funding possibilities are limited; there are no clear-cut economic and legal systems for organising production; industrial complexes are too large and highly specialised in products for which market demand has sharply decreased; there is no entrepreneurial tradition.
There are some industrial sites with good infrastructure as well as partly or fully completed industrial premises and structures.	Information on them is diffuse and poorly distributed, marketing is limited, the legislative base of privatisation is definitely imperfect.
There is great potential for the development of attractive new zones for housing construction.	There are occasionally conflicts between the development of the transport infrastructure and datcha estates; existing urban development plans do not account for future needs.
There are attractive suburban amenities and noteworthy historical monuments in the corridor region that may provide options for tourism and recreation.	There is no clear strategy as to how these options can be implemented, the vast datcha estates that have sprung up recently are threatening the environment
There are sufficient resources for water supply	Surface waters are often polluted, ground waters are prone to pollution due to insufficient or inadequate purification facilities to combat in the first instance surface run-off.
There is a wide recognition of the necessity of structural change.	Material incentives by local authorities for attracting investment are insufficient, there is no knowledge or experience of modern management and control methods.

Main Development Centres

The main centres that should play an important role in the social and economic development of the TEC region are the cities of Brest, Baranovichi, Minsk, Borisov and Orsha. The TEC is the main planning axis of the republic. The above centres are located at the intersection points with other axes. They are defined as the main development centres together with a further 13 cities in the General Scheme of the spatial organisation of the republic. The extent of their social and economic influence is considerable (Figure 1).

All cities covered are multifunctional centres. Minsk as the capital and largest city in the country additionally boasts a unique industrial complex, well developed research and design facilities, pure science institutions, burgeoning centres of innovation and the main offices of the largest banks in the republic.

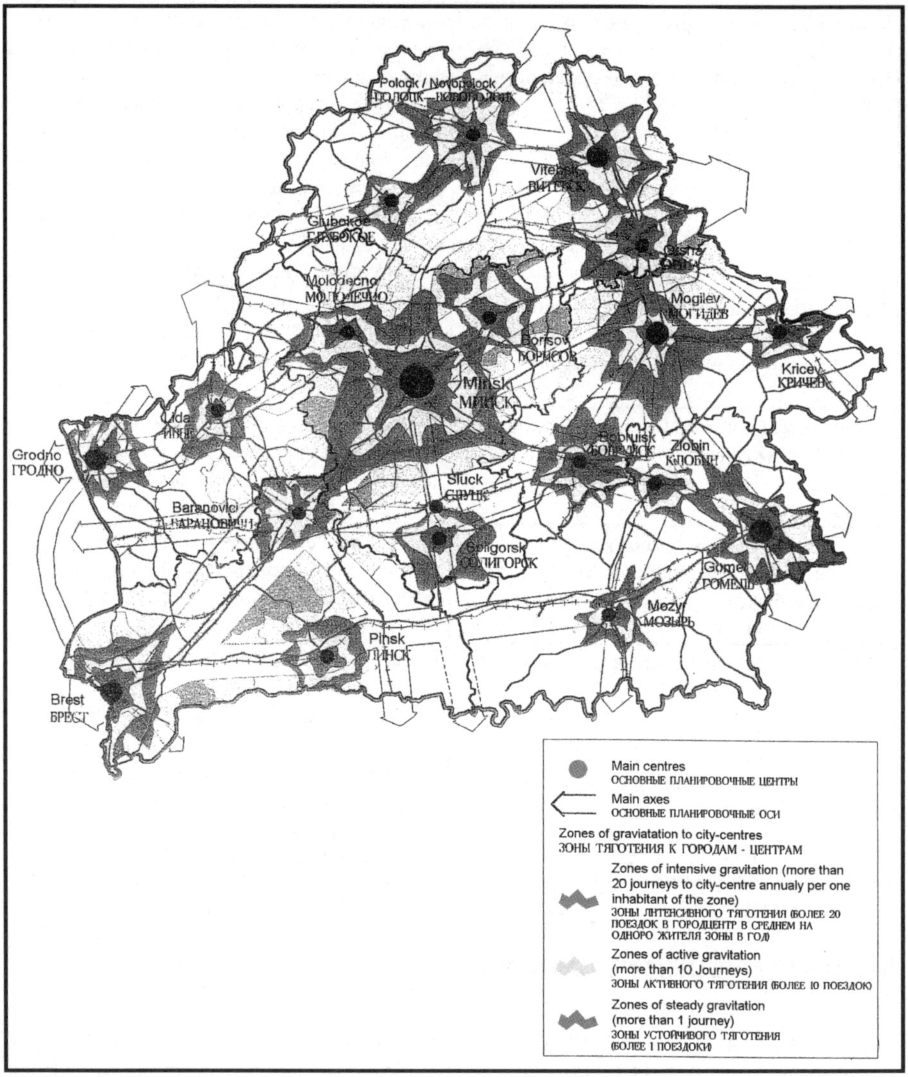

Figure 1. Spatial framework of Belarus

The development centres have a wide spectrum of prerequisites for further development: planning, social and economic, infrastructural (transport and utilities) and ecological.

All centres have, in course of their development, been integrated into the constantly growing transport and engineering infrastructure of the TEC (a network of trunk motorways, railways, gas pipelines, electric power supply lines, etc.) and have actively used it. Historically, transport links passing through cities determine the internal spatial organisation of these settlements.

Mainly owing to a favourable location within the planning structure of the republic, the outer zones of the above cities are host to a complex of transport facilities (from 2 to 4 types – motorway, railway, air, water) and utilities (high-voltage electric power lines, gas pipelines, oil pipelines) designed to attract new hi-tech manufacturing facilities, terminals, service centres, as well as facilitating the organisation of free economic zones.

Closely related to the active industrial development of cities is the sphere of personnel training and scientific services for the production sector. The availability of multi-profile secondary and tertiary special educational establishments makes it possible to establish new production facilities, enhance levels of qualification, and master new skills. As a result, the above cities even now possess highly qualified workers and engineers, and the educational level of the population of working age is higher than the average for the republic. The specific role of Minsk in this sphere is beyond any doubt; she is home to 55 % of the nation's professors, teachers and staff at higher educational establishments and 75 % of scientific staff.

Reserve capacities at cargo terminals on motorways and especially in railway traffic are useful preconditions for further developing the economic potential of the region and its main development centres.

A further factor favouring progress in the development centres of the region are the rich natural resources located nearby: attractive landscapes and recreational and nature conservation amenities. The availability, assuming rational use, of sufficient reserves of underground (including mineral and medicinal) and surface waters means the problem of supplying dwelling and industrial zones in these cities with water is solvable.

The main problem affecting the spatial development of cities is the absence of the legislative base needed to coordinate the development of cities and adjoining districts.

This explains the problem associated with the rezoning of land for the spatial development of cities, which has become very acute in recent years owing to changing patterns of housing construction. Increasing the share of low-rise construction with land plots (up to 15 % of total housing construction in Minsk and 30 % and more in other development centres of the region) is impossible without access to new areas of considerable size that are not available within municipal boundaries.

In all cities there is an increased demand for land for small businesses, a category not catered for in previous urban construction schemes.

Under the influence of changes in economic conditions, separate self-sustaining units will be set up within the majority of large enterprises. Restructuring of industry and general changes in expectations concerning the quality of accommodation will require new approaches to the way the planning of urban areas is organised.

In all the cities or in their immediate vicinity are to be found areas from which the armed forces have withdrawn, and these are, as a rule, well equipped with technical services and are hence potentially attractive for investors. However, their integration into the planning structure of the cities is difficult due to the unsolved status (at state level) of the above areas and the unwillingness of local authorities to take on their maintenance and development. These problems are most acute in the cities of Brest, Baranovichi and Borisov.

Due to the unfamiliar and unusual nature and great public importance of these tracts of land , complex consideration of territorial and urban construction and the economic aspects of conversion might, in our opinion, become a special topic for international cooperation,.

The concept takes into account future problems of urban development related to the planned construction of a high speed railway (HSR) in the region. The present task involves establishing its route. One should consider two aspects of the problem:

- optimal alignment of an HSR line;
- siting of passenger and freight stations along the HSR route. They are envisaged in all development centres. Their location should take into account the local transport infrastructure with the aim of organising quick and comfortable links between the main users of HSR services (city centres, residential districts, industrial sites) and also to earmarked sites outside municipal industrial centres (terminals, free economic zones, etc.).

Developing the Transport Infrastructure

The trans-European transport links in the region form the bedrock of the Belarus transport system and an integral part of the main transport network in Europe. On the whole, the quality of transport services in this region is higher than in other parts of the republic.

However, the M 1/E 30 highway does not meet European motorway norms. This is due to insufficient traffic safety, the poor quality of the surfacing, low permissible top speeds (100 km/h for motor cars and 80 km/h for trucks as well as speed limits of 60 km/h in built-up areas) and the poor quality of services.

Traffic density along the M 1/E 30 motorway varies between 4.3 to 8.9 thousand physical units per day. It is forecast to grow by 5–7 % per year. The bulk of trucks on the motorway run between Russia and Poland, Germany and other European countries. Motor car movements are predominantly domestic in nature.

Priority with respect to the M 1/E 30 motorway is currently being accorded the following tasks: improving the quality of the road surface; improving road markings and road signs (routing system); erecting safety barriers in built-up areas; installing technical and medical aid stations along the motorway.

Improving the quality and safety of traffic will facilitate elimination of the remaining, chaotically designed single-level intersections.

It is planned to turn the E 30 motorway into a toll road for all types and categories of road transport. Parallel roads are to be available for non-charge use. This procedure will also allow access restrictions for low-speed traffic to be introduced.

The problem of providing parallel roads may be solved either by rebuilding parts of the M 1/E 30 alongside the old road and transferring the older sections to local use, or else by adding non-toll lanes to the E 30.

Parts of the M 1/E 30 motorway still pass through built-up areas. As a result, the city transport infrastructure, especially in Brest, is being choked by through traffic. The problem of building bypasses in Brest is closely interlinked with the problem of building and upgrading border crossing points at the Belorussian-Polish border.

A considerable planning effect will be achieved by the building *of a southern motorway bypass for Brest.* It is taking shape in stages. The first stage is a 15-kilometre stretch of road along the southern perimeter of the city. For the longer-term future, there is an option to build a new cross-border motorway taking in Kobrin, Malorita, and Domachevo.

This option would also relieve the Kobrin-to-Brest section of the M 1/E 30 motorway and, consequently, divert transit traffic away from single-level intersections in built-up areas there.

The E 20 railway is a double-track electrified line. The general state of track-side installations and the permanent way are good. Passenger volumes on the E 20 are showing steady growth, but cargo shipments have halved over the past five years. As a result of the latter, freight yards and cargo handling equipment now have substantial reserves of capacity.

For the period 2000–2003, a line upgrade is planned to increase max. train speeds to 160–180 km/h, with the possibility of increasing the theoretical maximum speed to 200 km/h. It will be necessary in the process to replace all railway crossings with a system of regular bus services routed via overpasses.

At some sections it may be necessary to increase curve radii, modify the longitudinal section, and redesign the approaches to stations. This will involve special restructuring schemes needing to be implemented in concert with a municipality's general development plans.

For further coordinated development of the E 20 railway and the network of communities it is necessary to set land aside for the most important activities, including:

- additional (third) main lines at the main sections of the Minsk network (of highest priority are the Minsk–Zaslavl and Minsk–Rudensk sections followed by Minsk–Borisov and Minsk–Stolbtsy)
- creation of terminal and warehouse areas on railway premises at Minsk, Brest, Baranovichi and Orsha.
- refurbishment of passenger facilities at Brest–Tsentralnii station.

High Speed Railway Line

The planned high speed railway (HSR) between Berlin, Warsaw, Brest, Minsk and Orsha should ensure the integration of Poland, Belarus and Russia into the Trans-European Network (TEN) of high speed passenger links now being set up. The regional development concept takes into account that the consequences of building an HSR are quite considerable and that, in terms of economic impact, they may be compared to those associated with the construction of the Moscow-to-Warsaw railway in the 19th century.

Technical solutions as to the track gauge of the HSR (1,520 or 1,435 mm), the rolling stock, and the time-schedule for construction are currently at the discussion stage.

In 1992–1993 Russian and Belorussian experts considered the following options:

- increasing the maximum speed on the existing Brest–Minsk–Moscow railway line to 160–180 km/h while preserving the existing profile of the line (for the period up to the year 2000);
- increasing the maximum speed on the existing line up to 250 km/h with reconstruction of the line profile;
- construction of a special high-speed long-distance railway line for maximum speeds of 300 km/h and featuring a 1,520 mm gauge;
- construction of a high-speed long-distance railway line of 1,435 mm gauge (European standard).

At the same time, if a decision is made to use rolling stock with automatically adjustable running gear, the planning and economic aspects of this or that choice of HSR profile will become determinant.

Undoubtedly, the HSR will provide an opportunity to establish domestic high-speed passenger links between the cities of Brest, Baranovichi, Minsk, Borisov, Orsha.

Taking into account the realistic prospect of also using the HSR for high-speed freight traffic, it is very important at the initial stage to determine the location of possible interchanges between the HSR and existing railway networks and motorways. The project defines the land that should be reserved for freight terminals, sorting stations, and new industrial and residential areas.

Construction of such a railway is a very complex planning, land-use, urban planning and ecological task.

Traffic speeds of 300–350 km/h along the new line will require high smoothness of the track, with curve radii of over 7,000 metres and total isolation from adjacent land.

Laying the HSR alongside the new Brest–Minsk–Moscow motorway (E 30) is considered to be the most promising option. And even though the planning documents take into account the possibility of laying the line within areas of municipal jurisdiction, the details of proposals to allocate land to the HSR are insufficient.

New legally binding plans are necessary to coordinate the use of land, especially under the conditions of growing demands for land privatisation, mass multi-storey housing construction and the earmarking of land for new allotments and datcha estates.

Without waiting for the final decision on the HSR line's technical specifications, it is necessary at any rate to ensure the early availability of the land on which the railway is to be built. It is planned that the HSR route chosen shall eventually have the status of a zone of special state regulation.

Environmental Protection

Developing the transport infrastructure inevitably has consequences for the environment. Under the terms of the project, all land in the region is given a level-of - conflict rating from 1 to 5 to indicate the state of conservation as well as the likelihood of being able to utilise natural land of particular interest.

The concept of sustainable ecological development for the region is based on the following principles:

- in the process of improving the transport infrastructure and hence communications it is necessary to ensure healthy and safe conditions for human habitation are established;
- the creation of a communications corridor should be carried out on the basis of rational use and conservation of natural resources;
- environmental protection should be regarded as an integral part of the process of developing the communications system;
- the technical solutions adopted should in the first instance be designed to eliminate the causes of negative impacts on the environment, as opposed to simply softening the blow.

The above principles make it possible to define the main thrust of the catalogue of activities to be elaborated for the sphere of environmental protection and the rational use of natural resources:

- To minimise pollution of the atmosphere, soil, vegetation or water through emissions from road and rail traffic by:
  - introducing strict monitoring of the technical condition of vehicles along the whole of the motorway;
  - constructing a system of devices to confine pollution (including establishment of a system of protective woodland belts).
- To minimise negative impacts on the environment arising from motorway services by:

  - equipping premises with high-quality sewage facilities and installing a full technical infrastructure;
  - devising a reliable system for utilisation of solid waste;
  - locating services so as to take due account of sanitary norms relating to protection measures for drinking water.

- To uphold the diversity of the biological world and of landscapes and make rational use of land, woodland, and water resources by:
  - strict observance of the type of protection required by unique and valuable natural entities and complexes;
  - establishing the wherewithal to ensure the possibility of migration for the genetic stock of animals and plants;
  - applying construction methods that avoid hydrological damage to land, i.e. the flooding or desiccation of natural vegetation areas;
  - minimising the width of links where they traverse woodland, valuable agricultural lands, and in the suburbs of cities, where there is even a shortage of land for the development of residential estates;
  - equipping motorway services with recirculating water supply systems (where water is drawn from surface sources in districts with a deficit of surface water resources).

Implementation of the conceptual provisions for environmental protection calls for the elaboration of complex and sectoral programmes devoted to protecting the atmosphere, soils, surface and ground waters, communities and species of plants and animals, and to utilising waste. In the first place it is necessary:

- to limit economic activity in the vicinity of transport links within the boundaries of strictly protected areas;
- to construct a system of tunnels/bridges and to install special barriers for unhampered animal migration;
- to locate road services outside the boundaries of protected land;
- to establish an integrated system of forest protection belts;
- to use construction technologies which will not impair the hydrological characteristics of the area.

Development of Services along M 1/E 30 Motorway

The existing system of roadside services in the region does not comply with modern requirements either in terms of sophistication or location.

Owing to this, the competitiveness of the transport infrastructure is diminished and potential economic and social effects, i.e. income and employment growth, remain unrealised.

The objective prerequisites for development may be divided into three groups: technical, planning and investment.

The technical prerequisites for enhancing services are defined by the findings of an analysis carried out in 1995 as well as on the basis of special field and polling surveys performed in the summer and autumn of 1996; they include such parameters as traffic density, seasonal, weekly and hourly irregularity of transport flows, their structure by type of vehicle, aims and geography of freight and passenger trips, and the dynamics of motor car ownership in the local population.

Planning prerequisites define distances between service areas and their location relative to the main settlements, particularly protected and recreational areas. The distance between services fluctuates as follows: motels – from 1 to 180 km; petrol stations – from 2 to 65 km.

Investment prerequisites were defined on the basis of field surveys of roadside services as well as of data for land allocated by the local authorities for the planned construction of such services.

In recent years the process of locating and building services has been considerably advanced. The construction of three hotels at the Brest–Minsk section is nearing completion, while two service complexes with hotels at the Minsk-Orsha section are currently being built.

A considerable proportion of roadside services to be upgraded or newly built are presently undergoing approval tests by the local authorities as well as by architectural and construction bodies.

The general principles behind the upgrading and expansion of existing services and the location of new ones are as follows:

- normative distances recommended by the Agreement on International Highways and by TEM (trans-European motorways);
- symmetrical location of like services on both sides of the M 1/E 30 motorway in the interests of traffic safety. Location of services on one side of the motorway only is possible if there are overpasses nearby ensuring vehicles can access services without having to negotiate oncoming traffic.
- location of new services near settlements so as to derive the greatest advantage from labour resources and utility services by taking account of the prospects for future spatial development of these settlements as set forth in their General Plans;
- taking into account the specifics of location of the zones of special state regulation, including nature reserves, recreational zones, items of historical and cultural heritage (especially when locating motels and hotels).
- considering potential locations for customs and terminal-warehouse premises.

What is called for if, from a national interest point of view, the programme for the enhancement of services is to be implemented efficiently are changes in the existing practice of land allocation and approval procedures for the construction of services, together with reallocation of the functions of the public agencies, ministries and local self-government agencies involved in making the decisions on the location of new services.

Recreation and Tourism

The improvement of transport communications within the zone of the Berlin–Warsaw–Minsk trans-European corridor creates favourable prerequisites for the development of recreational amenities. Furthermore, the corridor region has great potential for the development of tourism since there are historical and cultural items of note and picturesque nature reserves and scenery.

Within existing recreation zones there is the capacity to locate facilities for long-term recreation for a quarter of a million visitors and more. Existing establishments are currently posting a utilisation factor of less than 15 %. There is also a well developed network of long-term recreational facilities for children.

The main goals of the recreation system in the region are as follows:

- revitalisation of the local economy; establishment of enterprises and agencies providing services for tourists; creation of new jobs; achievement of conditions for efficient use of natural, recreational and tourism resources.

Within the active impact zone of the transport corridor it seems feasible to develop the following main types of recreational activity:

- business tourism, short-term and long-term convalescence, cultural and instructional trips and excursions, and sport-related and recuperative tourism.

The efficient use of the region's cultural potential as a means of consolidating the tourism system is being hampered by a lack of public funds earmarked for research, design and construction works relating to the rehabilitation of architectural monuments in tourism zones and to the establishment of the requisite system of tourism services. The rate of privatisation in this sphere is low, as is the volume of private investment. Further factors include:

- poor information and low levels of local and international marketing;
- imponderables of state policy in the sphere of tourism development. The recently created Ministry of Sports and Tourism is only just beginning to draw up a tourism development concept. There are no clear legislative acts and provisions assigning special status to areas possessing valuable tourist and recreational resources or regulating procedures for the use of such resources;
- the low rate of privatisation and an underdeveloped market in recreation and tourism services. There is no entrepreneurial tradition. There is a need to conduct professional training for managers and service personnel taking into account the requirements and forecast inflow of local and foreign tourists;
- the low purchasing power of the population. The overwhelming majority of the population have low incomes. So far, only a small section of the population has been able to afford diversified recreation and tourism. Recreation and convalescence measures for children are mainly subsidised by the state budget and the trade unions.

An assessment of existing tourism potential and of the problems hindering its full exploitation gives rise to the identification of a set of organisational, legislative and planning initiatives designed to inject the necessary developmental momentum into this sphere.

In the period between the late 1970s and the early 1990s, allotment and datcha construction was fervently pursued in the corridor region, as it was in the republic as a whole. The biggest boom in allotment gardening occurred between 1989 and 1994, with the land involved almost doubling. As a result, by 1995–1996 about 35 % of citizens in large cities had their own separate garden plots on the outskirts. Indeed, the number of allotments is continuing to rise despite reductions in applications for rezoning for garden and datcha purposes and the fact that, on the property market, the supply of allotments considerably exceeds demand.

15.6 thousand ha of land in the corridor impact zone was given over to datchas at the beginning of 1995. On this land, there were 183.2 thousand plots with an average area of 0.07 ha. From 1991–1996 the total number of allotments within the corridor region increased by a factor of no less than 1.7.

Many datcha estates are directly adjacent to the lands allocated for the main railways and motorways. In suburban Minsk such a situation obtains at the Stolbtsy–Dzerzhinsk–Volcovichi and Sadovyi–Smolevichi–Zhodino–Borisov sections. This considerably aggravates the problems associated with the upgrading and even more so the building of transport links within this zone.

Since the majority of datcha plots are in private ownership, alignment of any new route, especially of an HSR, will involve considerable outlay as owners are bought out. The fact that new allotments are still being created makes it a matter of particular urgency that land be definitively allocated for the future HSR and assigned the status of a special state regulation zone in the sphere of architectural and civil engineering activities.

Where the building and upgrading of transport links is concerned, it would appear reasonable to lend priority to the following:

- defining the time-schedule of land reservation necessary to draw up design documentation and allocate land;
- prohibiting land allocation on reserved areas for allotment and datcha construction as well as any refurbishment of existing buildings;
- developing a regulatory and legislative base for the appropriation of land currently given over to datchas and of other real estate required for various purposes (building and upgrading of transport links, new enterprises, service facilities for tourism and recreation, and the social and technical infrastructures), including valuation procedures and mechanisms for compensating the owners, lessees and users of plots of land and the real estate thereupon.

Serious scientific and planning problems requiring special analysis are forecasts of growing demand for new allotments, the link between this demand and the needs in such areas for low-rise housing, and the options and conditions for the conversion of allotments and datcha estates into full residential districts.

Organisational Prerequisites for Active Regional Development

Economic development is possible given an ongoing improvement in the legislative machinery and the option of pursuing an active regional policy with direct public investment and other funding programmes.

Organisation of regional economic development presupposes an institution to which any potential investor may refer and which is capable of providing general and more specific local information on the possibilities and conditions for capital investment in a variety of sectors, industrial sites, and uncompleted or unused industrial premises; it should be capable of offering reliable solutions and thus of influencing the urban planning process to the greater good of the city.

This is why the project is proposing the establishment of regional and municipal economic promotion agencies (RPA) within local administrations, both for the corridor region and for other priority development zones in the republic. As is common practice in countries with a developed market economy, such agencies should become one of the instruments for the implementation of active regional policy. As a rule, an RPA acts as an intermediary between those responsible for making political decisions and those representing economic interests.

Forming part of a city or regional administration, an RPA is the instrument for solving all these problems. RPAs have the task of promoting economic development by supervising projects that need to be implemented in a short time so as to deliver marketing of the real estate and thus impact on urban planning with the aim of ensuring a sustainable process of development.

The sphere of influence of any large centre extends beyond municipal boundaries. That is why the RPA should have the authority to coordinate the needs and plans of both cities and their adjacent districts. It is very important and necessary to involve (through the channels of an RPA) all significant and influential organisations responsible for economic development in the decision making process.

It is for this reason that the RPA would need to have quick access to information meeting the needs and requirements of potential investors and entrepreneurs. This means availability of information concerning construction sites and buildings, specific economic services provided, services provided by RPA, living conditions in a city or district, and special opportunities to obtain financial and other support.

Regarding the need to perform these tasks without undue outlay, an RPA will be most efficient if it combines the functions of urban planning, economic consults, a statistical agency and a public relations centre. Focused and accurate single-source information on the conditions for economic development will allow prompt and reliable decisions to be made.

The identification, testing, and implementation of new and effective forms of organising spatial planning and management within a framework of drawn-out economic reform is a crucial prerequisite for sustainable development in all eastern European countries and can clearly be regarded as one of the main directions in which European cooperation needs to go.

Towards a Framework for Sustainable Spatial Development in Central and Eastern Europe

Ulrich Graute
Institut für ökologische Raumentwicklung (IÖR)
Institute of Ecological and Regional Development
Dresden, Germany

The more complex a problem is, the more unlikely it is a simple answer will be found for it. The problems surrounding sustainable spatial development in central and eastern Europe are exceedingly complex. Parallel to European integration and globalisation, the interdependence between territorial entities, between various sectoral policies impacting on the territory as a whole, and between various levels of government sharing responsibility for a particular territorial unit is growing. What makes spatial development so important in this context is the fact that all these trends take on tangible shape in the space in which we live and that it is here that the integrated approach to sustainable development actually has to be put into practice. Economic, social and environmental developments might have diverse origins but they all leave their footprints on the surface of the earth. The way we use land provides a complex picture of the economic, social and environmental conditions in which we live. Archaeologists can tell a lot about the lives of ancient people just by studying the ruins of houses, villages and cities. Spatial planners look more towards the future and ask how our cities and regions will need to look if sustainability is to be possible. In this context the outline and discussion of visions for spatial development and of strategies for their implementation have priority.

For central and eastern Europe as a whole a specific vision for spatial development and specific strategies for their implementation have yet to be developed. Nonetheless, promising work has been done to outline visions and strategies for subregions such as the Baltic Sea, whilst at the European level the debate on a European Spatial Development Perspective (ESDP) has got going.

The ESDP process foresees members from central and eastern European (CEE) countries participating in its further elaboration. This is important if a spatial development perspective is to be developed that is truly European and not just applicable to the EU. The moment ESDP were to become official EU policy, it would be the pre-eminent European policy on spatial development simply by virtue of the fact that it applies to 15 countries. Accordingly, it is not only interesting for eastern experts to participate in the debate, it is important that they do so.

In spite of existing EU initiatives and programmes and despite talk of their being improved and extended, the EU's ability to support the process towards sustainable spatial development in CEE will remain limited. Countries such as Hungary, Poland, and the Czech Republic that are readying themselves for EU membership may be better placed to profit from funding than others, but it is imperative that all countries eventually up their own input and elaborate their own approaches to spatial development and related transnational cooperation. In the following, three priority fields of activity are set out, to wit:

- *public and political debate in CEE;*
- *infrastructure for spatial development cooperation in CEE;*
- *spatial research with a European perspective.*

1 Public and Political Debate on Spatial Development in Central and Eastern Europe

Public debates take time, yet given the necessarily longer-term perspectives of spatial development that time is certainly available. Towards the end of her contribution (page 59), Karolijn van Ginneken has the following to say of spatial development in CEE: "First of all, even countries that are used to working together, with similar planning cultures and a similar political framework, have needed three years to reach a consensus on a common spatial planning agenda. It is a gradual process and you cannot force your way through it. It takes time." It is improbable that a group of experts from CEE countries could develop a CEE spatial development scheme within a short period of time that would meet with general approval from the public and government. Conditions within the various countries, the vested interests of the various groups of players, and levels of cooperation across CEE are too divergent. Countries such as Poland and the Baltic States are already vigorous participants in VASAB 2010 cooperative ventures on spatial development around the Baltic (cf. Zaucha) while within the Commonwealth of Independent States tension is still running so high that not even a catastrophe like Chernobyl proved motivation enough for countries to coordinate joint initiatives with a longer-term dimension (see contributions on Chernobyl). Evidently, it will take time to build up cooperation not only in western but also in eastern countries. Furthermore, debate and cooperation that take place exclusively amongst planning experts run the danger of being misconstrued in CEE as a continuation of the old system of central planning as it existed in Soviet times. To increase acceptance of cooperation in spatial development, it will be necessary to initiate public and political debates in CEE, firstly on the general objectives and tasks of spatial development planning in democratic societies and, secondly, on the specific need for such a coordinated development in CEE.

The conferences of the Network of Spatial Research Institutes in Central and Eastern Europe at Vilnius (1995) and at Odessa (1996) revealed clearly the disparity between points of departure in transnational cooperation. At Vilnius it was possible to discuss a vision and strategies for the Baltic region that had been outlined a year before the event, while at the conference on the Black Sea the very fact that the participating government representatives from Ukraine expressed a willingness to enter into cooperative arrangements at some point in the future had to be counted as a success. Deep cooperation around the Black Sea is still the exception. Only where coastal zone management is concerned have the countries in question embarked on meaningful cooperation, an instance that is all the more remarkable for being the exception (cf. Kharitonov). A two-pronged public and political debate could serve to increase awareness of the various stages of development involved as well as emphasising the importance of coordinated spatial development at regional, national, and transnational level.

2 The Infrastructure for Cooperation on Spatial Development in Central and Eastern Europe

While public and political debate on transnational cooperation in spatial development is only just getting off the ground in most CEE countries, some cooperation is actually already possible using existing instruments and infrastructures. CEE countries bordering the Baltic have got together in the Committee on Spatial Development in the Baltic Sea Region (CSD–BSR), spatial research institutes are cooperating in a purpose designed network (*CEE Net*), and international funds from the European Union and elsewhere are also providing opportunities for transnational cooperation on spatial development.

2.1 Cooperation at Governmental Level

The example of the Committee on Spatial Development in the Baltic Sea Region

Since 1992, representatives from national and regional ministries responsible for spatial planning and development in the Baltic Sea Region have been pooling efforts towards producing a framework of spatial development for the region. They jointly prepared and published a document 'Vision and Strategies around the Baltic Sea 2010 (VASAB 2010)', which was approved by the Third Conference of Ministers of Spatial Planning and Development at Tallinn (December 7–8, 1994). At the same conference, a Committee on Spatial Development in the Baltic Sea Region (CSD–BSR) was established. Through conferences of ministers, the committee of government representatives, the supportive VASAB 2010 secretariat in Gdansk, pilot projects and common initiatives, the Baltic Sea Region has taken a lead in transnational cooperation between politicians and planners (for more information see Selke, Zaucha and Fischer).

2.2 Cooperation Between Governmental and Non-Governmental Players

The example of the Network of Spatial Research Institutes in Central and Eastern Europe (CEE Net)

During preparations for the VASAB 2010 document, it became apparent that there was a need to rationalise the knowledge-gathering process with regard to both the factors shaping spatial development and the instruments and options available for enacting spatial policies. Accordingly, a resolution of the Tallinn conference stressed the importance of promoting spatial research and the dissemination of scientific advice to the appropriate authorities. At this point, what had hitherto been international cooperation between governments in the Baltic Sea Region became transnational cooperation between government and non-government organisations. The notion that transnational cooperation in spatial planning involves a need for consultation in elaborating concepts as well as improved legal and financial instruments and the training of specialists was not new however.

The German Federal Ministry for Regional Planning, Construction and Urban Development is supporting central and eastern European countries in their process of transition from centrally planned economies to economic and social entities meeting the requirements of market economies. To this end, the ministry took the step of inviting representatives of research institutes (i.e. of the existing research infrastructure in CEE countries) to a seminar at Leipzig on 21–22 January 1993. The objective cited by the ministry was the notion of improving the pooling of information on current research projects and findings in the belief that intensified cooperation within a network would help enhance the quality of the consultation process where spatial development is concerned. It was envisaged that the network would enable institutes to initiate cooperative research projects and to benefit from research findings from other institutes. A list of the potential goals and tasks of the network was compiled during discussions at the first network seminar in Leipzig:

- forum for meetings and exchanges of information on institutes' activities and research findings;
- identification of research topics of common interest;
- promotion of research on the social and economic effects of transition in eastern Europe;
- harmonisation of standards and methods (statistics, maps, spatial monitoring systems, research methods);
- harmonisation of terminology, dictionary of planning terms;
- discussion on ecological and social standards for spatial planning;
- support for central and eastern European countries in the development of new planning legislation;
- transfer of know-how from west to east (joint research projects, seminars,

working visits);
- common research on border areas;
- improvement of basic tools, methods and information for spatial research and policy advice;
- response to an increasing need for policy advice in view of ongoing integration and internationalisation;
- identification of and joint planning for critical regions of European significance;
- evaluation of experiences with different spatial planning systems in European countries;
- support for eastern European planners in their efforts to promote spatial planning in a period of rapid economic and political change.

Among those actively involved in the network is the *Institut für ökologische Raumentwicklung* (Institute of Ecological and Regional Development, IÖR), Dresden. Since 1995, the institute in Dresden has also been doubling up as the network's secretariat. This body has a remit to act as an information centre for research and research institutes, distribute information in the form of a newsletter, and organise conventions such as the annual network conferences. As cooperation between those party to the network has intensified, so interest in publishing findings in works such as the present book has grown.

As is to be expected, the input of network participants varies considerably. Since *CEE Net* is an open network with no formal membership, it is even difficult to stipulate the exact number of those party to it. The mailing list includes more than 200 addresses and is still growing, but it would be a somewhat complex task drawing a definitive line between 'active' and 'passive' contributors. In each country there are from one to five institutes strongly interested in participating at network meetings. *CEE Net* is one of the biggest networks of its kind, has interlocutors in all parts of the region, is interrelated with other networks and already has five years of cooperation under its belt. Nonetheless, the role of a network like *CEE Net* is limited. It is able to cope with dynamic situations and to act where informal cooperation is useful or necessary, but without major funding and a more formal structure it cannot provide the necessary infrastructure for lasting research cooperation.

3 Spatial Research with a European Perspective

Given the limits on funding, differing political priorities in the various CEE countries, and the fact that spatial issues are accorded varying importance across CEE, efforts embracing all CEE countries and requiring national funding currently offer little promise of being implemented. It will remain unrealistic for some years yet to attempt to draw up a comprehensive spatial development perspective for CEE. A more promising route will be to support cross-border, transnational and interregional cooperation wherever possible and to follow and, where necessary, to improve the example of the Baltic Sea. In the Stockholm Declaration on Sustainable Spatial Development Policy in the Baltic Sea Region of Ministers for Spatial Planning (22.9.1996), it was concluded that, 'Eventually, regional development projects will have to be realised in line with the economic, social, and ecological criteria of sustainability' and that the European Commission should establish the necessary preconditions to ensure that joint development initiatives and projects on both sides of the EU's common external border can be financed out of funds provided for under the Community Initiative INTERREG II C and, for countries in transition, the support programmes PHARE and TACIS. As long as regional cooperation such as that in place in the Baltic region and that gradually taking shape in the Danubian and the Black Sea regions is based on the same principles and guidelines as are to apply to cooperation within EU (cf. ESDP, IV.A), then it will be possible in a subsequent phase to integrate these regional development modules into a pan-European scheme.

A lot of research and development can be conducted at regional and national level, and regional development initiatives will greatly facilitate the drafting of a pan-European development scheme. This in turn will necessitate a pan-European spatial research institution. A specification of European visions for spatial development will not be possible without specific data and information. Comparable analytical data and information on spatial development observed over longer periods of time can only be provided by a European spatial research institution liaising closely with research institutes in the European countries. At the informal council of ministers in Leipzig (1994) and in the draft of the European Spatial Development Perspective, there had already been proposals to establish a European 'observatory'. Such an observatory would presumably resemble a research institute, but the question then needs to be asked as to whether the observatory's function would really just be to observe spatial development at European level, collate the relevant data, and then simply leave scientific analysis to national institutions. Admittedly, it is important to have national and indeed regional institutes at which spatial development in general is studied from a national or regional perspective, but by extrapolation it would be just as important to have a European institution which does the same from a European point of view.

The example of the Baltic Sea and the other instances of cooperation on spatial development discussed in this publication indicate that reaching consensus on the basic thrust and overall goals of spatial development appears to be a fairly straightforward affair. But what these examples also demonstrate is that it is a lot more difficult to tackle the subsequent far more involved stages in transnational spatial development. Who shall be responsible for implementing, in which way and with the support of whom, an overall goal such as that enshrined in the principles of sustainability as formulated in the Agenda 21 document from the Rio Conference and as habitually cited since in discussions on spatial development for all levels? A top-down planning process and a centralised research infrastructure would not be viable means of achieving sustainable spatial development, but the fact that in central and eastern Europe and at European level there is no leading institution or organisation whatsoever that could at least coordinate cooperation in spatial development is one of the reasons why efforts in this direction have so far only yielded limited success.

Developing a framework for sustainable spatial development in central and eastern Europe involves more than simply outlining a development scheme. What is also needed, and to an extent should actually be placed 'up-stream' of other activities, are moves to initiate a public and political debate, to train specialists, and to set up or refine infrastructures for cooperation on spatial research and development in central and eastern Europe and at European level.

Subject Index

Contributors

Altzinger, Wilfried, Dr.
Vienna University of Economics and Business Administration,
Augasse 2–6, 1090 Vienna
Austria

Anderson, Vladimir N., Prof.
Department of Economic and Social Geography
Odessa State University, GIS Laboratory,
ul. Petra Velikogo 2, 270000 Odessa
Ukraine

Bassi, Tarcisio
Deputy Director of Environment and Local Authorities (DELA)
Council of Europe,
67075 Strasbourg CEDEX
France

Fidrmuc, Jarko
Institute of Advanced Studies,
Stumpergasse 56, 1060 Vienna
Austria

Fischer, Helle
Ministry of Environment and Energy
Spatial Planning Department,
Højbro Plads 4, 1200 Kopenhagen K
Denmark

Graute, Ulrich
Institut für ökologische Raumentwicklung
Institute of Ecological and Regional Development,
Weberplatz 1, 01217 Dresden
Germany

Gukalova, Irina, Dr.
Institute of Geography, National Academy of Sciences of Ukraine,
ul. Volodymyrska 44, 252003 Kiyv (Kiev)
Ukraine

Horlenko, Inga, Dr.
Institute of Geography, National Academy of Sciences of Ukraine,
ul. Volodymyrska 44, 252003 Kiyv (Kiev)
Ukraine

Ivlichev, Victor, Dr.
Director of the Institute for Regional and Urban Planning of Belarus,
ul. Varvasheni 81, 220002 Minsk
Belarus

Kharitonov, Igor A., Dr.
International Activity Centre on Development of Common Methodologies
for Integrated Coastal Zone Management (BSEP-ICZM/IAC),
ul. Krasnaja 19, 350063 Krasnodar
Russia

Khazdan, Isaac, Dr.
Institute for Regional and Urban Planning of Belarus,
ul. Varvasheni 81, 220002 Minsk
Belarus

Kozlovskaya, Ludmila, Dr.
Belarussian State University,
ul. Skoryna 4, 220050 Minsk
Belarus

Malchus, Viktor Freiherr v., Dr.
Former President of the Academy of Spatial Research and Regional Planning,
Grabengasse 47 (Am Eulenturm), 59423 Unna
Germany

Maier, Gunther, Dr.
Vienna University of Economics and Business Administration,
Augasse 2–6, 1090 Vienna
Austria

Pascariu, Simona
URBANPROIECT Institute,
Str. Nicolae Filipescu Nr. 53–55, 70136 Bucuresti Sector 2
Romania

Roch, Isolde, Dr.
Institut für ökologische Raumentwicklung
Institute of Ecological and Regional Development,
Weberplatz 1, 01217 Dresden
Germany

Rudenko, Leonid, Dr.
Director of the Institute of Geography,
National Academy of Sciences of Ukraine,
ul. Volodymyrska 44, 252003 Kiyv (Kiev)
Ukraine

Schneidewind, Peter
Director of the Austrian Institute for Regional Studies and Spatial Planning
Österreichisches Institut für Raumplanung (ÖIR),
Franz-Josefs-Kai 27, 1010 Wien
Austria

Selke, Welf, Dr.
Head of the European Spatial Planning Division
Federal Ministry for Regional Planning, Construction and Urban Development,
Deichmanns Aue, 53179 Bonn
Germany

Shilov, Alexander S., Prof.
Russian Academy for Public Service under the President of the Russian Federation
Faculty of Social Ecology,
ul. Vernadski 84, Moscow
Russia

van Ginneken, Karolijn M., Drs.
Ministry of Housing, Spatial Planning and Environment
National Spatial Planning Agency
Rijksplanologische Dienst
IPC 352,
Rijnstraat 8, 2515 XP Den Haag
The Netherlands

Vartiainen, Perttu, Prof.
Department of Geography
University of Joensuu,
P. O. Box 111, 80101 Joensuu
Finland

Yatsukhno, Valentin, Dr.
Belarussian State University,
ul. Skoryna 4, 220050 Minsk
Belarus

Zaucha, Jacek, Dr.
Deputy Secretary of VASAB 2010,
Dlugi Targ, 80828 Gdansk
Poland

Printing: Mercedesdruck, Berlin
Binding: Buchbinderei Lüderitz & Bauer, Berlin